LANCHESTER LIBRARY, Coventry University
Gosford Street, Coventry CVI 5DD Telephone 024 7688 7555

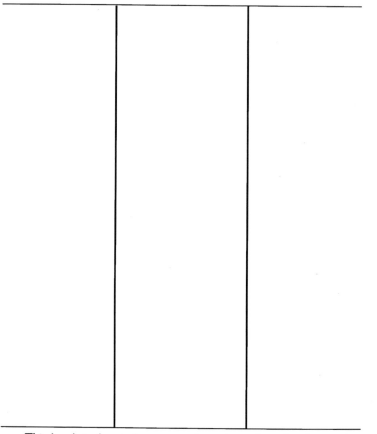

This book is due to be returned not later than the date and
time stamped above. Fines are charged on overdue books

DIETARY SUPPLEMENTS
AND HEALTH

Novartis Foundation Symposium 282

DIETARY SUPPLEMENTS AND HEALTH

BICENTENNIAL
1807
WILEY
2007
BICENTENNIAL

John Wiley & Sons, Ltd

Other Wiley Editorial Offices

John Wiley & Sons Inc., 111 River Street, Hoboken, NJ 07030, USA

Jossey-Bass, 989 Market Street, San Francisco, CA 94103-1741, USA

Wiley-VCH Verlag GmbH, Boschstr. 12, D-69469 Weinheim, Germany

John Wiley & Sons Australia Ltd, 33 Park Road, Milton, Queensland 4064, Australia

John Wiley & Sons (Asia) Pte Ltd, 2 Clementi Loop # 02-01, Jin Xing Distripark, Singapore
129809

John Wiley & Sons Canada Ltd, 6045 Freemont Blvd, Mississauga, Ontario L5R 4J3, Canada

Wiley also publishes its books in a variety of electronic formats. Some content that appears in
print may not be available in electronic books.

Novartis Foundation Symposium 282

viii + 228 pages, 19 figures, 15 tables

Anniversary Logo Design: Richard J. Pacifico

British Library Cataloguing in Publication Data

A catalogue record for this book is available from the British Library

ISBN: 978-0-470-03427-9

Contents

Participants

Peter J. Aggett Lancashire School of Health & Postgraduate Medicine, University of Central Lancashire, Preston PR1 2HE, UK

Jan Alexander Norwegian Institute of Public Health, PO Box 4404 Nydalen, N-0403 Oslo, Norway

Angelo Azzi Vascular Biology Laboratory, Room 622, JM ISDA-HNRCA at Tufts University, 711 Washington Street, Boston, MA 02111, USA

Alan R. Boobis Experimental Medicine & Toxicology Section, Division of Medicine, Imperial College London, 9S10B Commonwealth Building, Hammersmith Campus, London W12 0NN, UK

Kevin D. Cashman Department of Food and Nutritional Sciences, University College Cork, Cork, Ireland

Paul M. Coates National Institutes of Health, Office of Dietary Supplements, 6100 Executive Blvd., Room 3B01, Bethesda, MD 20892-7517, USA

Edzard Ernst Institute of Health and Social Care Research, Peninsula Medical School, Universities of Exeter & Plymouth, 25 Victoria Park Road, Exeter EX2 4NT, UK

Peter Gann Department of Pathology, University of Illinois at Chicago, 1819 W. Polk Street, CMW 446 M/C 847, Chicago, IL 60612-7335, USA

Barry Halliwell (*Chair*) Department of Biochemistry, Yong Loo Lin School of Medicine, National University of Singapore, MD-11 Clinical Research Centre, 01-10, 10 Medical Drive, Singapore 117597

Peter Hollman RIKILT and Wageningen University, P.O. Box 230, 6700 AE Wageningen, The Netherlands

Martijn B. Katan Faculty of Earth & Life Sciences, Vrijye Universiteit, De Boelelaan 1085, 1081 Amsterdam, The Netherlands

Marguerite Klein National Center for Complementary and Alternative Medicine, 31 Center Drive NCCAM, Bethesda, MD 20892-208, USA

Claudine Manach INRA-U3MR1019, INRA Clermont/TheixUnité Nutrition Humaine, Centre de Clermont-Ferrand/Theix, 63122 St Genès Champanelle, France

Yesim Negis (*Novartis Foundation Bursar*) Marmara University, Faculty of Medicine, Department of Biochemistry, 34668 Haydarpasa, Istanbul, Turkey

Hildegard Przyrembel Federal Institute for Risk Assessment, Thielallee 88–92, Berlin, D-14195, Germany

Margaret Rayman Division of Nutrition, Dietetics and Food Science, School of Biomedical and Molecular Sciences, University of Surrey, Guildford GU2 7XH, UK

Robert M. Russell Jean Mayer USDA Human Nutrition Research Center on Aging (HNRCA), Tufts University, 711 Washington Street, Boston, MA 02111-1524, USA

John M. Scott School of Biochemistry and Immunology, Trinity College Dublin, College Green, Dublin 2, Ireland

Paul Shekelle RAND Evidence-Based Practice Center, 1776 Main Street, P.O. Box 2138, Santa Monica, CA 90407-2138, USA

Roland Stocker Centre for Vascular Research, Bosch Institute, University of Sydney, Medical Foundation Bldg K25, 92–94 Parramatta Road, Camperdown, NSW 2006, Australia

Christine Lewis Taylor HF-33 FDA, Office of Science and Health Coordination, 5600 Fishers Lane, Rockville, MD 20857, USA

Brian Wharton MRC Childhood Nutrition Research Centre, Institute of Child Health, 30 Guilford Street, London WC1N 1EH, UK

Elizabeth A. Yetley National Institutes of Health, Office of Dietary Supplements, 6100 Executive Blvd., Room 3B01, Bethesda, MD 20892-7517, USA

Eu Leong Yong Department of Obstetrics & Gynaecology, National University Hospital, Yong Loo Lin School of Medicine, National University of Singapore, Lower Kent Ridge Road, Republic of Singapore 119074

Chair's introduction

Barry Halliwell

Department of Biochemistry, Yong Loo Lin School of Medicine, National University of Singapore, MD-11 Clinical Research Centre, 01-10, 10 Medical Drive, Singapore 117597

The supplements industry is huge, worth billions of dollars each year. The products marketed range from things that are known to be essential in the human diet (such as vitamins C and E, and folic acid) to those that some people claim are essential, but the evidence is equivocal (such as chromium and vanadium), and to non-nutrients, such as flavonoids, which are increasingly mixed with vitamins in supplement tablets as well as used to fortify foods. There is the whole field of traditional Chinese medicine (TCM), which encompasses a wide area, but the important part in the context of this meeting is the use of herbal remedies. This is very important in Asia, but is also growing across the rest of the world. Since I was last in London another two Chinese medicine halls have appeared in my local area, taking the number from three to five.

For some of these supplements (e.g. calcium, folic acid), the regulatory authorities have already decided that health claims can be made and food can be fortified. All areas of the world are trying to come to grips with the best ways of regulating and controlling the supplements industry. Equally problematic is the question of how one attempts to set estimated average requirements and recommended dietary allowances (RDAs) for nutrients known to be essential to the human diet, on the basis of an inadequate database. One thing we would like to address at this meeting is to suggest what research should be done to enable us to build our knowledge base of what is really required in the human diet; what methods should be used for this research (biomarkers, human studies, animal studies and so on) and how valid these methods are; and who should do the work (for example, should it be funded by industry?).

I was at a meeting recently where a distinguished speaker made the comment that all the published positive results on health benefits of vitamins E and C have been paid for by the manufacturers. I don't know whether this is true or not, but it is an indication of how much of the research in this field is supported by industries that have a vested interest. This doesn't mean that it is bad research or invalid research. Certainly, I found when I worked in the UK, apart from the Ministry of Agriculture, the granting agencies were not particularly interested in human nutritional research.

It is not only nutrients that are essential in the human diet that are widely sold. For example, there is an increasing marketing of fortified foods with increased levels of flavonoids. For the papers that will be presented at this meeting, we have picked certain topics. In some cases this is because a lot of work has already been done (for example, for vitamin E, we know an enormous amount about its mechanisms of action and its effects on human disease), and this history might inform us about what to do or what not to do with the other nutrients. That said, if someone asks me after all this research what is the RDA for vitamin E, there are different figures for the US, UK and Asia, and I'm not convinced by any of these figures. I would say we don't really know. A lot of work has also been done on folates, vitamin D and selenium (a fascinating nutrient because there are suggestions that some parts of the world are deficient and need more, while people in other areas may risk excess; it's a nutrient where one can easily overdose). Finally, we should look at the requirements for herbal remedies. How should these be regulated? They are becoming increasingly popular. Sometimes they are impure or adulterated, or they are the wrong plant. If we regulate by saying they are intrinsically unsafe until proven otherwise, the whole industry may be driven underground.

So we shall look at vitamin E, folic acid, vitamin D, flavonoids, selenium and traditional Chinese medicine as case studies. How much do we know, and how can we draw conclusions and apply them to a wider network? This isn't really a meeting to discuss regulation, although some of our conclusions may have implications for regulatory authorities. Our goal here is to ask what scientific evidence we would need before we would be convinced that a supplement was safe, and to ask how we could gather that evidence: who will do that work and who will pay for it? The key part of this meeting is not the presentations, but the discussions, and I look forward to our interactions over the next few days.

Risk assessment of dietary supplements

Alan R. Boobis

Section of Experimental Medicine and Toxicology, Division of Medicine, Imperial College London, Hammersmith Campus, Ducane Road, London W12 0NN, UK

Abstract. Risk assessment of dietary supplements shares many of the requirements of that for other chemicals, although there are some important differences. Amongst these is the essential nature of some nutrients so that it may be necessary to balance the need to minimize toxicological risk with the need to avoid deficiency. There may also be limitations on experimental design, in that high doses may not be achievable for nutritional reasons and available human data on toxicological hazard is likely to be very limited. Prior to embarking on a risk assessment the problem needs to be formulated. This involves risk assessors, risk managers and relevant stakeholders. A key decision is whether a risk assessment is necessary and, if so, what is required of the assessment. This will shape the nature and output of the assessment. Risk assessment itself is a scientific process comprising four steps, hazard identification, hazard characterization, exposure assessment and risk characterization. Hazard identification involves determining the range of toxicological effects that might be caused by the substance, whilst hazard characterization establishes dose–response relationships, toxicological and species relevance of the findings and establishes health based guidance values. Exposure assessment involves predicting or measuring the level, pattern and duration of intake of the substance by exposed individuals. This may require dietary consumption data. Finally, risk characterization is the process whereby all of the prior information is integrated to reach conclusions in a form appropriate to the question posed. The nature of the output can take several different forms, and may be qualitative or quantitative. There are some cross-cutting issues in risk assessment, primarily on uncertainty and variability. The sources of uncertainty at each step of the risk assessment should be clearly identified and quantified to the extent possible. Variability requires that the risk assessment should take into account all relevant subpopulations and groups, where there might be differences on the basis of intake or sensitivity. This would very often include different life stages, but may also include gender, ethnic or genetic differences. The report of the risk assessment should be systematic and transparent, identifying all key assumptions and defaults used.

2007 Dietary supplements and health. Wiley, Chichester (Novartis Foundation Symposium 282) p 3–28

Dietary supplements have been defined as products that contain one or more ingredients, such as vitamins, minerals and herbs, that are intended to supplement the diet, are intended for human use and are in the form of a tablet, capsule, powder or another preparation that is not a conventional food. Dietary supplements also include other botanicals (excluding tobacco), amino acids, fibre, essential fatty acids, dietary substances to supplement the diet by increasing the total daily intake (e.g. enzymes or tissues from organs or glands), concentrates such as meal replacements or energy bars, metabolites, constituents, or extracts. According to EU Directive 2002/46/EC, only those supplements that have been proven to be safe may be sold without prescription. Hence, in order to market a supplement within the EU, it is necessary to provide evidence of a reasonable expectation of safety in normal use (European Parliament 2002).

Dietary supplements can be grouped according to their chemical or biological characteristics. Chemically, a supplement can be a single well-defined chemical that is produced either synthetically or naturally. In such cases, the chemical identity and purity should be specified. Supplements can also be mixtures of chemicals, usually from natural sources, either of defined composition or as complex mixtures, the exact composition of which is unknown. In this case, there should be evidence of batch-to-batch consistency and any contaminants need to be below acceptable maximum values. Biologically, supplements could be considered essential, in that some intake is necessary for normal health, or non-essential. Essential nutrients are those substances (elements and compounds) that are either not made in the body or are made at insufficient levels to maintain adequate concentrations, which are uniquely necessary for normal intermediary biochemical processes involved in ensuring structural and functional integrity of the organism. In general, it is not possible to replace one essential substance with another. A key aspect of essentiality is the role of homeostatic mechanisms, which regulate the levels in the body. An essential nutrient may have biological properties in addition to those associated with its essentiality, often at intakes above those necessary to maintain sufficiency. A non-essential nutrient serves some function in intermediary biochemical processes that is not unique to the substance itself and hence it can be substituted by another with similar properties. At high levels of exposure, both essential and non-essential nutrients may have additional biological properties, which could be considered more pharmacological than nutritional.

Risk assessment

The risk assessment of dietary supplements, particularly those comprising essential nutrients, poses particular complexities, because of the need to avoid deficiency on the one hand and toxicity on the other. On a population basis, no threshold can be identified for the dose–response curve, as this follows a normal distribution. It is always possible that someone, somewhere is exquisitely sensitive. Pragmatically, a threshold for exposure below which there is no appreciable risk can be established (see below). Similarly, in theory at least, it is not possible to identify

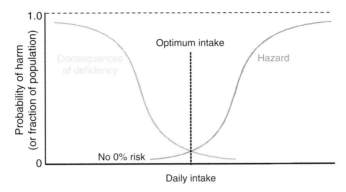

FIG. 1. Intake–response relationships for dietary supplements of potential adverse health effects that may arise at low intakes as a consequence of deficiency (left hand curve) and at higher intakes due to a toxicological effect (right hand curve). Providing that the nature of deficiency and the nature of toxicity are of similar adversity, the optimum intake occurs where the curves overlap (see text). Note that the curves are cumulative probability distributions and hence, in theory at least, do not reach the x axis.

a dose above which everyone is sufficient. The consequence is that, as shown in Fig. 1, as intake decreases, risk of toxicity will decrease whereas the risk of deficiency will increase and vice versa, and there will be no absolutely clear separation between the two curves. However, when the doses associated with very low respective risks are well separated, it is not difficult to make recommendations but when these doses are close to each other or even overlap, it may be necessary for the risk assessor to provide information on several different scenarios, to help the risk manager in reaching a decision (Renwick et al 2004). One scenario often addressed is that dose which minimises the risk of toxicity and the risk of deficiency. This occurs where the curves overlap, and has been described as the optimum intake. However, the concern here is that only the probability of an effect is considered, with no consideration of severity. In the absence of some method of comparing these quantitatively, the risk manager may wish information on the relative risks from different exposure scenarios.

Problem formulation

Prior to embarking upon a risk assessment it is necessary to formulate the problem that is to be addressed. During this activity preliminary to the risk assessment, consideration needs to be given as to whether a risk assessment is required, who should be involved in the risk assessment and subsequently in the risk management and how the output from the risk assessment will provide the necessary support for risk management (Renwick et al 2003). Problem formulation involves dialogue among risk assessors, risk managers and other relevant stakeholders such as representatives from the food industry, consumers and health professionals. As such, it establishes the basis for the risk assessment rather than comprising a step in risk assessment.

Problem formulation commences with identification of a potential issue. This may come about by any of a number of ways, including development of novel products requiring assessment or authorisation, health surveillance studies, including epidemiological studies, identifying a potential concern, advances in scientific knowledge requiring a reconsideration of existing exposures and consumer concerns. During problem formulation a decision must be taken as to whether there is a need for a risk assessment. Risk assessments are usually resource and labour intensive, and hence should not be undertaken lightly. Once the decision to conduct a risk assessment has been taken, it is important that the necessary resources are made available to ensure that the process is timely and effective.

In reaching a decision as to whether a risk assessment is required, a variety of factors should be taken into account. These include prior knowledge of the substance and any previous use:

Prior knowledge on the substance

- Origin of substance
- History of use and consumption
- Chemical identity, characterization and specification
- Effect of processing on substance and on whole food

Prior knowledge on exposure to the substance

- Prior knowledge on possible biological effect(s)
- Qualitative aspects
- Quantitative aspects
- Predicted effects

On the basis of this information, consideration needs to be given as to whether there is any human (systemic) exposure and whether there is a potential hazard. In the absence of information, the possibility that there might be a hazard is often assumed. If there is no exposure and/or no potential hazard, no risk assessment is necessary. If there is both exposure and potential hazard, a risk assessment is necessary. However, it may still not be feasible to conduct an assessment, for example because of the minimal data available. Hence, only if an assessment is both necessary and feasible should it proceed.

Once such a decision is taken, agreement will be needed on who should be involved in the risk assessment and in the risk management processes, how the risk assessment will provide the information necessary to support the risk management decision, the adequacy of the database in supporting a risk assessment, whether the necessary resources are available and the timeline for completing the risk assessment. All available information will need to be collected prior to the risk assessment. In addition to full details of prior knowledge, information will be needed on specific (sub)populations that are to be a focus for the risk assessment, including geographical location and consumption patterns, relevant route(s) of exposure, though this is not a major issue for dietary supplements, and possibly the health endpoints that are to be considered.

It is important that the outputs of the risk assessment, i.e. the risk characterization, are appropriate to the problem that was identified initially. In addressing the requirements of risk management of dietary supplements, these should be clearly specified, understood and clarified through dialogue, as necessary.

The risk assessment paradigm

Risk assessment has been defined as a process intended to calculate or estimate the risk to a given target organism, system, or (sub)population, including the identification of attendant uncertainties, following exposure to a particular agent, taking into account the inherent characteristics of the agent of concern as well as the characteristics of the specific target system. It is a scientific exercise comprising four steps: hazard identification, hazard characterization, exposure assessment and risk characterization (Fig. 2). Whilst this paradigm was first developed for xenobiotic chemicals it is also applicable to dietary supplements.

Toxicokinetics

Information on the kinetics of the substance, including data on absorption, distribution, metabolism and excretion, can be invaluable in helping design and/or interpret toxicological studies in experimental animals and studies in humans (both interventional and observational) (Barton et al 2006). Absorption of nutrients occurs in the gastro-intestinal tract and may be by passive or active processes. Nutrients are often subject to active transport, under homeostatic regulation (Said & Mohammed 2006). The form in which the supplement is consumed can affect the extent of absorption, for example through complexation or binding (Gibson et al 2006). Some compounds are subject to presystemic metabolism by intestinal microflora or by enzymes in enteroyctes or hepatocytes. This could result in reduced systemic bioavailability. Hence, the systemic bioavailability, which can be estimated from the area under the plasma concentration time curve (AUC), relative to the AUC after intravenous administration, is influenced both by the extent of gastro-intestinal absorption and the extent of presystemic metabolism.

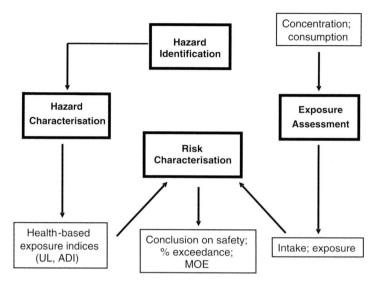

FIG. 2.   The risk assessment paradigm. Risk assessment is a scientific process comprising four steps: hazard identification, hazard characterization (which includes dose–response assessment), exposure assessment and risk characterization. Whilst hazard identification and hazard characterization are shown as distinct steps, in practice there is often considerable overlap between them. Some of the inputs and outputs of the various steps are illustrated.

Following absorption, substances are distributed throughout the body. Concentrations in tissue compartments, such as the brain, testes and the fetus, are determined by the physicochemical properties of the agent and whether it is subject to active influx or efflux transport. As part of their homeostatic regulation, many nutrients are bound to plasma proteins and are transported across various cell membranes and tissue barriers. Such processes exhibit specificity and may be saturable, both of which properties may vary with species and life stage, for example.

Many compounds undergo metabolism that reduces or eliminates their biological activity. In addition to the enzymes of intermediary metabolism, which may play a role in the biotransformation of nutrients, the enzymes of xenobiotic metabolism may be involved, depending on the structure of the substance. Xenobiotic metabolism has been divided into two phases. Phase I involves functionalization of the molecule whilst phase II involves conjugation with an endogenous donor moiety. The enzyme systems most involved in such metabolism are the cytochromes P450 in phase I and the sulfotransferases and glucuronyltransferases in phase II. These enzymes are widely distributed throughout the body, but are often most active in the liver and small intestine. Differences in the level of expression and the specificity of the enzymes of xenobiotic metabolism are a common source of biological variability in the toxicological response to foreign compounds (Barton et al 2006, Singh 2006).

The parent compound and any metabolites will be excreted from the body at a rate that is dependent on their physicochemical properties. For example, highly lipid soluble compounds will generally be eliminated more slowly than very water soluble compounds. The main routes of excretion are via the kidneys in the urine and via the liver in the bile. Passive diffusion plays an important role in such excretion. However, it is now apparent that there are a number of active transport systems involved in both renal and biliary excretion (Shitara et al 2006). As such they will vary in their specificity and activity, for example between species.

All processes that depend on specific binding, whether enzymatic metabolism or transport, are potentially saturable as well as discriminating towards structure, although in the latter case specificity can be quite broad. The consequence of these properties is that kinetics may change with dose, due to saturation of the various processes involved and there is the possibility for interactions between compounds that share a binding site. Lipid soluble compounds may also be excreted in breast milk, resulting in exposure of suckling infants.

The rate of elimination, whether by metabolism or excretion will determine the extent to which a compound will accumulate in the body, relative to the frequency of exposure, and how rapidly the compound will accumulate to a steady state.

It is often possible to obtain information on the kinetics of dietary supplements by studies *in vivo* in human volunteers, as well as in experimental animals. In addition, many of the processes involved in the kinetic behaviour of dietary supplements are particularly amenable to studies *in vitro*. For example, it may be possible to obtain comparative data between experimental animals and humans in this way.

Hazard identification

Hazard has been defined as the inherent property of a nutrient or other substance to cause adverse health effects depending upon the level of intake. It should be noted that elsewhere hazard has sometimes been defined in terms of the agent itself, but this is not consistent with the risk assessment paradigm described here. For a nutrient, any adverse effects associated with excess intake will be a consequence of it inherent properties, but the risk of deficiency cannot be considered to be due to any inherent properties of the nutrient. Hence, some alternative definition of hazard is necessary for this unique situation. To date, no such definition has been agreed.

Hazard identification involves identification of the range of adverse affects that can be caused by an agent, at all relevant life stages (Cooper et al 2006). It is important that due consideration be given to duration of exposure, to cover all likely human exposure scenarios, and that the range of studies is sufficient to detect all potential endpoints of concern. Hazard identification also involves a conclusion regarding the relevance or otherwise of the effects observed for human health. Some effects may be dismissed as not being adverse whilst others may be dismissed as they are species specific and not relevant to humans. Obviously, any such decision must be soundly based and presented in a transparent manner.

Hazard identification is usually based on data from either observational or experimental studies in humans or from experimental studies in laboratory species. It may also be possible to obtain relevant data from studies *in vitro* (Barlow et al 2002, Eisenbrand et al 2002, van den Brandt et al 2002). Whilst data from controlled clinical trials is potentially the most useful for risk assessment, the type of information obtained is often more on nutritional effects; there are ethical and practical difficulties in obtaining data for most toxicological endpoints. Whilst these concerns are overcome to some extent in observational studies, these are limited by the quality of the intake data, which is often not very reliable or accurate.

Studies in experimental animals can encompass a wide range of endpoints and all life stages. However, in the case of dietary supplements care must be taken that the high doses that can sometimes be achieved do not cause non-specific nutritional changes, for example by altering caloric or total protein intake. Inter-species extrapolation can also be of concern for some dietary factors, due to differences in nutritional requirements.

Few *in vitro* or *in silico* methods have yet reached the stage where they can provide reliable data on chemical hazards. However, these approaches can be invaluable in providing mechanistic information, in studies of species differences and as part of a tiered approach to hazard evaluation (Eisenbrand et al 2002). The use of any non-validated method should be accompanied by adequate data on its performance characteristics and fitness for purpose.

As indicated above, potential hazards to all life stages need to be identified. Whilst this can be problematical in human studies, in experimental animals it is possible to evaluate effects throughout the developmental period and indeed on successive generations (Cooper et al 2006). Whilst the different stages of development in experimental animals do not necessarily map precisely to the equivalent stages in humans, by covering the entire period of development it should be possible to obtain an holistic assessment of potential life stage-specific hazards.

Whilst many hazards can be identified following relatively short periods of exposure, in the order of weeks or months, there are certain hazards that only become apparent after prolonged exposure, in the order of 1–2 years in rodents. Hence, in studies of hazard identification it is important to consider the possible exposure duration in humans. For example, is the supplement likely to be taken only for a relatively short period of time or is lifetime intake possible/probable?

In assessing the available evidence for hazards that might form the basis of the risk assessment, it is important to evaluate each effect for toxicological and human relevance. Effects of concern in risk assessment, i.e. hazards, are those that are adverse. An adverse effect has been defined as a change in morphology, physiology, growth, development, reproduction or life span of an organism, system, or (sub)population that results in an impairment of functional capacity, an impairment of the capacity to compensate for additional stress, or an increase in susceptibility to other influences. In this context, some effects may not be adverse, particularly some of those produced by essential nutrients, where the normal processes of homeostatic regulation may operate. Changes within the range, and as a consequence, of homeostatic regulation are not normally considered adverse within this definition.

Over the last few years, there has been much interest in identifying biomarkers of adverse effects, which may help in determining the adversity of the response to a chemical or other agent (Gundert-Remy et al 2005). Such biomarkers often reflect an intermediate process in the causal pathway to toxicity that is necessary but not sufficient for the adverse effect. A biomarker may also reflect a process that is associated with, but not necessary for, the adverse effect. An example of the former would be a change in creatinine clearance, indicative of deteriorating renal function, whilst an example of the latter would be inhibition of red cell acetylcholinesterase, itself not adverse but indicative of potential effects on the neuronal enzyme. The development of biomarkers is being assisted by the ability to assess globally changes in protein or metabolite levels, by proteomics and metabonomics respectively. However, whilst these can support discovery of candidate biomarkers, much work still has to be done to determine the specificity and sensitivity of the candidate biomarker, and whether it is fit for purpose.

For dietary supplements, there will often be situations where it is not possible to establish an unequivocal association in humans between a biological response, as a putative biomarker of effect, and an adverse health outcome. However, by combining information from different types of studies it may still be possible to establish such a link. For example, it may be possible to establish such a relationship in experimental animals and to use *in vitro* data obtained using animal and human derived systems to provide a bridge between species, thereby demonstrating a causal relationship between exposure, the biomarker of effect and a toxic effect. Hence, no single system provides sufficient data to enable validation of the biomarker, but taken together, data from several systems may provide weight of evidence for the validity of the biomarker (IPCS 2006).

Hazard characterization

Once the potential hazards of a substance have been identified the next step in the risk assessment process is hazard characterization, which includes dose–response assessment (Dybing et al 2002). Whilst hazard identification and hazard characterization are normally portrayed as two separate, sequential steps in risk assessment, in practice there is often substantial overlap between them, so that hazard characterization is performed at the same time as hazard identification. A key component of hazard characterization is characterizing the relationship between the dose or intake and the incidence or severity of adverse effects. In addition to the shape of the dose/intake response curve, additional information sought is whether there is a threshold dose below which there is no response and if so at what dose/intake the threshold occurs.

The dose/intake–response curve may show either the incidence or severity of an effect. Incidence represents the number of individuals/experimental animals exhibiting a specific effect, for example offspring with cleft plate. Such data are said to be discrete. Severity is usually represented as a continuous variable, for example creatinine clearance. In the former, the number of 'responders' relative to the group size is plotted whereas in the latter, the numerical average and a measure of variability, e.g. SD or SEM, are plotted. In characterization of the dose/intake–response relationship information should be obtained on whether the response increases progressively with exposure, i.e. monotonically, or whether there are is some other relationship, such as U-shaped; the shape of the curve (for example, is it linear, log-linear, sigmoidal, hyperbolic, supra- or sub-linear?); how rapidly the response changes with exposure (steepness of curve); the maximum response (for continuous variables); and whether there is a threshold, and at what exposure this occurs (Fig. 3).

In addressing some of these questions, it is necessary to make inferences about the quantitative relationship between exposure and effect. In practice this can be achieved either visually or by curve fitting (dose–response modelling, DRM) (Edler et al 2002). Often, the approach that is used is determined by the reference point that is to serve as the basis of the risk assessment. The reference point is usually either the no observed adverse effect level (NOAEL), i.e. the highest dose or intake at which there is no significant response relative to the control, or the benchmark dose/intake, i.e. that dose or intake at which a defined low but detectable response (usually 5% for continuous data and 10% for discrete data), the benchmark response, occurs. This is abbreviated to the BMD or BMI (benchmark dose or intake). In practice, the lower confidence limit on the 95% confidence interval is used, to give the BMDL5 or BMDL10, respectively (Fig. 3).

There are advantages and disadvantages to both the NOAEL and the BMD approach. By definition, the NOAEL is a defined dose or intake. In experimental studies, it represents one of the dose levels used in the study. Hence, the NOAEL is critically dependent on dose spacing and variability in the data. The lowest dose/intake at which a significant response, relative to the control, occurs is defined as the lowest observed adverse effect level (LOAEL). The true no adverse

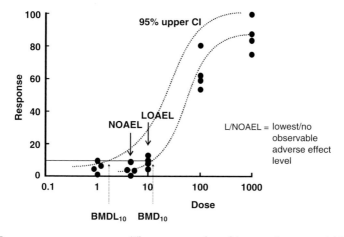

FIG. 3.   Dose–response assessment. The response plotted is a continuous variable, and individual data from each dose group are shown. The NOAEL is highest dose at which the response is not significantly different from the control. The LOAEL is lowest dose at which the response is significantly different from the control. The LOAEL and the NOAEL will be influenced by the choice of doses (dose spacing) and intra-group variability. Also shown is the result of dose-response modelling, where a best-fit curve is fitted to the data. A specific response is defined, the benchmark response, here 10% and the corresponding dose, the benchmark dose (here the BMD_{10}) is determined by interpolation. It is also possible to determined the upper 95% confidence interval for the curve and determine the corresponding lower limit on the BMD, here the $BMDL_{10}$. Note that the BMD and BMDL are not dependent on dose spacing and that the lower the variance in the model, the closer the BMDL will be to the BMD.

effect level will occur somewhere below the LOAEL, and may be above or below the NOAEL. The BMD(L) does not depend on dose spacing, as it is an interpolated value. It is obtained by fitting a mathematical relationship to the data, the curves often having no underlying biological implications (IPCS 2004). Goodness-of-fit criteria can be used to identify the best fitting model, but not infrequently the data can be described equally well by more than one curve and in such situations there may be no way to choose between alternative possibilities. The model is then used to obtain the 95% confidence interval and the lower limit used to determine the BMDL. Hence, the less variability in the data, the more precise the estimate of the BMDL. The BMD does rely on the number of dose/intake groups and requires that the difference in response between these groups covers a reasonable span of the overall dose/intake–response curve.

Much of risk assessment is based on the concept of a nominal threshold for a toxicological effect. Exposure below this threshold is considered to be without appreciable health risk. However there is much debate about whether toxicological responses exhibit true thresholds and if so how reliable estimates of these thresholds are. For most toxicological responses, at least at the level of the tissue or the organism, there is evidence that the processes involved exhibit a threshold. This is because of factors such as receptor response, rate-limiting networks and cascades in biochemical pathways, homeostatic regulation, both biochemical and physiological, adaptation and repair. However, there are a few toxicological effects which, at least in theory, might not exhibit a biological threshold. The most obvious of these is carcinogenicity arising from DNA-reactive genotoxicants. In such cases it is not possible to identify a NOAEL. In contrast, a BMD(L) can be determined, but its use in risk assessment differs from that for a response that is thought to exhibit a threshold (see below).

During hazard characterization, information on the time to onset of the toxicological response can be extremely valuable. For example, can the effect arise from a single exposure, or does the effect require prolonged exposure. A related issue is whether there are periods of vulnerability, particularly during development when relatively short term exposure will have an effect but only over the critical period. Another issue is whether effects observed are reversible, and over what period. This is of particular importance for exposures that are intermittent or episodic.

Whilst hazard identification and characterization can be undertaken empirically, the risk assessment can be enhanced considerably by knowledge of mode or mechanism of action. A mode of action can be defined as a series of empirically observable, precursor steps (key events) that are necessary elements of the mode of action, or are markers for such elements, which result in an adverse health effect (Sonich-Mullin et al 2001). This contrasts with a mechanism of action, which provides a detailed molecular description of the biological processes responsible for an adverse health effect. Identification and characterization of the key events in a mode of action provide the basis for a systematic and transparent consideration of whether a toxicological effect observed in experimental animals is relevant to humans, or can be discounted on the basis of fundamental qualitative differences in the key events or quantitative differences in toxicokinetics or dynamics between experimental animals and humans (Boobis et al 2006, Seed et al 2005). Even if the toxicological effect cannot be discounted, information on the key events can be invaluable in continuing the risk assessment, for example in assessing dose–response relationships, inter-species extrapolation and potentially vulnerable subpopulations.

Once the toxicological profile of an agent has been characterized, it is necessary to identify the critical adverse health effect. This is the most sensitive relevant endpoint in the most sensitive relevant species. If data of adequate quality are available in humans, this would take precedence over data obtained in experimental animals. However, it is important to consider whether the studies/observations in humans would have covered all endpoints of potential concern, including carcinogenicity and developmental toxicity. In practice the critical adverse health effect will be the toxicological effect with the lowest NOAEL or BMD(L) after excluding any effects that are clearly not relevant to humans or that are considered not adverse. This NOAEL or BMD(L) has been termed the reference point by some, as it will form the basis of the risk assessment at levels of human exposure. The suitability of the critical adverse health effect in protecting all potentially susceptible sub-populations needs to be considered, particularly different life stages. The corollary that also needs to be considered, is the critical adverse health effect overly protective of some sub-populations. An obvious example of this is an endpoint based on *in utero* exposure. Whilst relevant to pregnant women, this would clearly not be relevant to pre-pubertal children or to male adults. This can be of significance when separate data on exposure are available for such groups (see below), so that separate risk characterizations can be undertaken for the various groups.

Once the critical adverse health effect and its associated reference point have been identified, it is necessary to extrapolate to human exposures, taking into account uncertainty and variability in the human population, to derive a health based guidance value (IPCS 2006, Renwick et al 2004). For vitamins and minerals this has been termed the upper level (UL). For some other substances permitted in the diet this has been termed the allowable daily intake (ADI). Whilst, ideally, quantitative information on the magnitude of the various uncertainties and on interindividual variability would be used to derive a health based guidance value, in practice this is never available, or at least not completely. In the absence of specific information, default factors are used in such extrapolations. These have been termed uncertainty factors, but are also known as safety factors. The principal factors used are to allow for inter-species differences and interindividual variability. In non-nutrient risk assessment, these factors are both 10 by convention. They are assumed to be independent of each other, and the overall uncertainty factor is obtained from their product, i.e. 100. If human data are used, there is no need for an inter-species uncertainty factor and hence an overall factor of 10 would be used. For some nutrients, application of such an uncertainty factor can result in a health based guidance value (e.g. UL) that is close to or even below the level necessary to ensure nutritional adequacy, for example iron and zinc (Fig. 1). Hence, uncertainty factors for such substances need to be considered on a case-by-case basis, taking into account established intake requirements.

Where specific information is available in inter-species or interindividual differences in either toxicokinetics or toxicodynamics, obtained from studies *in vivo* or *in vitro*, it may be possible to derive chemical specific adjustment factors (CSAFs) (IPCS 2005). The default uncertainty factor of 10 for interspecies differences can be subdivided into sub-factors of 4 and 2.5 for toxicokinetic and toxicodynamic differences, respectively. Similarly, the default factor for interindividual differences can be subdivided into sub-factors of 3.16 and 3.16, respectively. Where specific information is available on a chemical, one or more of these sub-factors can be replaced by a data-derived value, the remaining sub-factors retaining their default values. The sub-factors are multiplied together to giver a composite uncertainty factor. Hence, the default uncertainty factor remains 100, unless modified by chemical-specific information.

Other uncertainty factors are used to allow for the use of a LOAEL rather than a NOAEL (or when it is not possible to determine a BMR considered to be toxicologically 'acceptable', see above), when the duration of the study is considered insufficient to fully characterize the hazard with respect to potential duration of human exposure, deficiencies in study design or in the database and lack of information relevant to potentially susceptible sub-groups. The magnitude of these factors is generally not greater than 10, and the actual value should be decided on a case-by-case basis. It is also important to consider whether any of these uncertainties is of such concern that it is not possible to complete the hazard characterization. In deriving the health based guidance value, all of the uncertainty factors are multiplied together.

The health based guidance value is obtained by dividing the reference point, usually the NOAEL or BMDL, by the overall uncertainty factor. This provides an intake value, expressed in mg per kg body weight per day, that is unlikely to lead to adverse health effects in humans.

For toxicological effects that do not have a threshold, or which in theory might not have a threshold and for which there is no evidence to the contrary, it is not considered appropriate to extrapolate to human exposure on the basis of uncertainty factors (Edler et al 2002). There are two basic approaches for dealing with such substances. The first is to conclude that no exposure level is acceptable and that, as a consequence, levels should be as low as reasonably practicable or achievable. If a substance requires market authorization the simplest way to achieve this is to withhold such authorization. Where this is not an option, either the dose/intake–response curve can be extrapolated to levels of human intake or an estimate of the margin of human intake over that producing a specified response can be calculated. In both cases, an estimate of likely human intake is necessary, and this is dealt with further in the section below on risk characterization.

Extrapolation from a reference point obtained from studies in experimental animals requires consideration of two types of threshold, which are illustrated in Fig. 4. The first is a biological threshold. This is identified in experimental animals from data obtained in the different dose groups. Variation between animals is considered uncertainty and the no adverse effect level, for which the NOAEL is a surrogate, can be considered a true threshold below which no response will be produced or detected, no matter how many animals are used or the sensitivity of

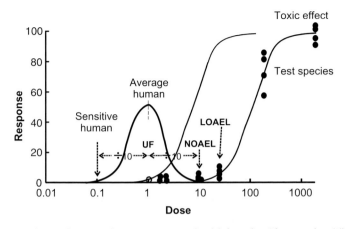

FIG. 4. Extrapolation from a reference point to a health based guidance value. The NOAEL is a surrogate for the biological threshold of the toxicological response observed in experimental animals. This has been described as the reference point. It is also possible to use a BMDL as a reference point. In the absence of evidence to the contrary, it is assumed that on average humans may be 10 times more sensitive than the test species. Hence an uncertainty factor (UF) of 10 is used to extrapolate to the putative biological threshold in an average human (O). However, humans will vary in their sensitivity, and this is illustrated as a Gaussian distribution of threshold values for the population. To allow for interindividual variation a default uncertainty factor of 10 is used. This is used to extrapolate to a nominal population threshold, which is defined as the health based guidance value. However, because there is a distribution of human sensitivities, there is a theoretical possibility that a very small number of individuals will be even more sensitive than assumed on the basis of the guidance value, i.e. in the leftward tail.

the measurement. The default 10-fold uncertainty factor for interspecies differences then extrapolates, on the default assumption that the average human may be up to 10-fold more sensitive than the most sensitive animal species, to an estimate of the biological threshold in the average human (Fig. 4). The uncertainty factor for interindividual variability is then used to extrapolate from this population mean to a population threshold, again using a default factor of 10. This is designed to protect an individual who is 10 times more sensitive than the average. However, if one considers that each individual has their own NOAEL (or reference point) then the population can be represented by a distribution of these values (Fig. 4). Hence, the population threshold, whilst protecting the vast majority of the population may not be completely protective of the most sensitive individual within the entire world's population. It is for this reason that risk assessment dose not assure absolute safety and that it is not practical or productive to seek to protect every individual, as there is no absolute threshold for a probability distribution such as that illustrated (Fig. 4).

Exposure assessment

The third stage of risk assessment is exposure or intake assessment. Exposure can either be estimated or measured, depending on whether a substance is being proposed for use or is already is use, respectively (Kroes et al 2002). A detailed description of exposure assessment is beyond the scope of this document, but a few key points are described here. Assessment of exposure to dietary supplements requires information on the levels of the supplement used by the consumer together with data on food consumption where this is relevant, for example when it is not possible to separate intake of the supplement from background exposure. In such cases it would be necessary to estimate total exposure and correct for the fractional contribution of the supplement to the total. A key issue is whether it is only intake of the supplement that is of concern in the risk assessment or whether it is total intake including that from the normal diet, for those substances that occur naturally as part of the diet. The latter requires detailed information on patterns of food consumption and composition.

It is not only the level of exposure that is of concern in exposure assessment. Information is required also on the frequency of exposure, how often is there any intake, and the duration of exposure, is the supplement such that exposure is likely to be infrequent or is it more likely to occur over a prolonged period of time. Are there identifiable sub-populations in whom the level or duration of exposure may be markedly greater than in the general population, for example on the basis of gender, life stage or ethnicity? Where a substance is not currently in use, exposure has to be estimated from likely use patterns, based on the intended function, use of similar products and predicted market penetrance. For those substances that are already in use, exposure can be estimated either from concentrations or levels permitted or from actual measurements in consumers. There are a number of approaches to obtain such information, such as questionnaires, diaries, household surveys and marketing data. There are advantages and disadvantages to all of the approaches in use. In general a balance needs to be reached between the resources required to conduct the study and the accuracy required, the greater the accuracy required the more resource intensive the study.

No estimate of exposure is truly accurate and hence there will always be some uncertainty in the data. The sources of uncertainty should be identified and to the extent possible quantified, at least in terms of whether they are likely to lead to under- or overestimation (EFSA 2006). These uncertainties should be transparently reported. Exposure also varies within the population, and sometimes between sub-populations. This variation needs to be captured in some way. Often this is achieved by presenting the data as a population distribution or a series of distributions, and specific percentiles of intake may then be used in the risk assessment, for example the 97.5th percentile. There may be several different distributions involved in the intake assessment, for example there may be data on consumptions of several different food items, together with levels of the substance of concern in these together with other sources of exposure. In such cases, a conservative approach is to take a high percentile from each distribution and then combine the data. A more realistic estimate can be obtained by undertaking a probabilistic assessment, in which the various distributions are combined to obtain a distribution of predicted or estimated intakes. Where intake data are available for different subpopulations this should be indicated, and the respective data used in the risk characterization (see below).

Risk characterization

The final step in the risk assessment is risk characterization, in which information from the other three steps is integrated to address the problem formulated at the outset and to provide output in a form that is most useful for risk management (Renwick et al 2003, 2004). A key component of the risk characterization is a comparison of the intake levels considered as unlikely to lead to adverse health effects in humans with predicted or actual levels of intake. At its simplest, if intake exceeds the health based guidance value, this would provide an indication that there might be some toxicological concern arising out of the use of the substance, and vice versa. Where toxicological and/or intake data are available for different subpopulations this conclusion can be extended to separate population groups. Thus, for example, it might be concluded that there is no concern in any population, including infants and women of child-bearing age. Alternatively, it might be concluded that there is exceedance of the health based guidance value only in a particular subpopulation.

Alternative outputs are possible from the risk characterization. Rather than extrapolate the reference point using uncertainty factors to obtain a health based guidance value, it is possible to calculate the margin of exposure (MOE). This is the ratio of the reference point for the critical adverse effect to the estimated or actual intake levels. This approach does not necessarily require that any assumption be made as to whether there is a threshold. When the BMD(L) is used as the reference point, it is equally applicable to effects considered to have a threshold and to those that are not. However, the interpretation of the MOE will differ. For those compounds considered to have a threshold, the same considerations as to uncertainty and variability will apply as when deriving a health based guidance value. Hence, exposures resulting in MOEs of greater than 100, when the only uncertainty is due to interspecies and interindividual differences, are considered unlikely to lead to adverse health effects in humans. Where there are additional sources of uncertainty, as described above, a greater MOE will be required before such a conclusion can be reached. For compounds that may not have a threshold in the dose/intake–response relationship, no MOE can be considered without risk but at large MOEs, the risk may be considered acceptable. From recent use, MOEs would need to be at least 10 000 before this would be the case (Barlow et al 2006). An alternative approach to dealing with such compounds is to extrapolate the dose/intake–response curve to predicted or actual levels of human exposure. This often involves linear extrapolation from the reference point, such as the BMDL. The output can be a slope factor, or risk per unit of exposure, expressed as mg per kg body weight per day. Alternatively, the risk as a probability or incidence of events per unit population can be provided for a given intake. Generally, risks of 10^{-5} or 10^{-6} are considered acceptable. This approach is more widely used in the USA than in Europe. It also has to be stressed that the need to undertake risk characterization for a substance with no threshold in the dose/intake–response curve is likely to be exceedingly rare for a dietary supplement.

The risk characterization should include a narrative that provides a clear and transparent consideration of the key issues involved in formulating advice to risk managers, together with an appraisal of the sources of uncertainty and the level of confidence in the conclusions (IPCS 2006, Renwick 2003, 2004). The nature of the adverse health effects produced by the substance should be described, together with any information on the mode or mechanism of action. Whilst the critical adverse health effect is of obvious importance, information should also be included on other possible effects, particularly in different life stages, where there may be differences in sensitivity so that although the critical effect in infants might drive the risk assessment, a less sensitive effect in adults may be the most critical in that group. Information on the severity of the effects and their reversibility can be of value. The nature of the dose/intake–response curve should be described, indicating whether or not it is monotonic, the steepness of the curve, whether a threshold is anticipated on biological grounds and if so where it is located, the nature of the estimate of the reference point (e.g. NOAEL, $BMDL_{10}$, LOAEL), with an explanation of why this particular type of reference point was chosen and any specific points to note. This could include the presence of potential outlying dose groups, wide dose spacing in parts of the dose range, close proximity of NOAEL to LAOEL, or particularly high variability in dose groups.

The uncertainty factors used in deriving health based guidance values should be reported and their choice explained as explicitly as possible. It is important that this information is transparent, as often during the risk assessment choices are made based on sound considerations but the reasons for these are not reported, making it difficult to understand later why a particular health based guidance value was established. The individual uncertainty factors should be listed and the composite factor used should be stated (the product of the individual factors). Any other factors used in the analysis should be explained and their numerical value provided. These might include corrections for bioavailability or species differences in body burden. The sufficiency of the data and the quality of the data should be reported. Any obvious data gaps, their potential impact on the assessment and whether they are considered critical should be described. Other sources of uncertainty should also be described, at least qualitatively and to the extent possible quantitatively. Where quantification is not possible, some indication of their impact on the assessment should be provided, if known, i.e. is the assessment likely to be less or more conservative on the basis of the uncertainty in question. Areas of uncertainty that should be specifically addressed include intake assessment and the assumptions necessary, whether the risk to all sections of the population has been adequately addressed, and uncertainty in the current state of knowledge, for example areas of biology where there is current disagreement over the significance of certain effects, the relevance or otherwise to humans of some toxicological effects or modes of action observed in experimental animals.

The basis for deriving the health based guidance value should be clearly stated. This includes identifying the specific study or studies from which the critical reference point has been obtained, the species and strain in which the effect(s) was observed, the duration of the study, the effect(s) observed at doses/intakes above the reference point, if any, if the reference point is the highest dose/intake for which there are data this should be stated, as should the basis and magnitude of the uncertainty factor used (see above). The population or sub-populations to whom the health based guidance value applies should be reported.

The basis for estimating intake should be described, including information on the methodology used, details of the populations surveyed, how representative the data are of all sub-populations of potential concern, characterization of intake, whether deterministic or probabilistic, explicit reporting of the percentiles used, if any, data on the average and high level intake, major assumptions and data gaps.

The predicted or actual exposure is then compared with the health based guidance value, for each population group of potential concern. Often, this will be based on life stage differences, for example infants, toddlers, children, and adults, who may differ in intake and/or in toxicological sensitivity (see above). The simplest conclusion from the risk characterization would be that intake was either above or not above the health based guidance value. Where there is exceedance, for those substances requiring authorisation prior to marketing, the choice would then be to either refine the risk characterization, perhaps by obtaining more accurate data on intake, or to decline approval. However, advice to risk managers arising from risk characterization could take a number of additional forms, depending on the problem that was posed. An alternative to a health based guidance value, the margin of exposure, is discussed above. In this case, the interpretation of the importance of uncertainty in the risk estimate is more the responsibility of the risk manager than the risk assessor, as a decision has to be taken as to what magnitude of the MOE would be considered unlikely to be of concern. Advice could be provided that, based on considerations of the different sources of uncertainty discussed above, intake resulting in an MOE above a certain value would be considered to pose a negligible risk, or would be unlikely to pose a risk. The values used to allow for uncertainty in arriving at this MOE would be similar to those used to calculate an overall uncertainty factor used to derive a health based guidance value as described above. As for health based guidance values, different MOEs could be used to assess the risk to different subpopulations, on the basis of differences of intake or in sensitivity to the toxicological effects. MOEs may be applicable to both threshold and non-threshold effects, but interpretation will differ, particularly as to what might be considered an 'acceptable' MOE (Barlow et al 2006). MOEs may also be used to rank and prioritize risks.

It may sometimes be appropriate to specify a maximum and minimum intake, such as would be the case for essential vitamins and minerals. However, this would require input from those with specific expertise in nutrition, to determine recommended minimum levels.

Other possible outputs from the risk characterization include estimates of the proportion of the population or of different subpopulations exceeding a health based guidance value, and by what magnitude. Such information could be provided both for those with average intakes and those with high intakes. It is also possible to highlight any subpopulations at particular risk, perhaps on the basis of genetic differences.

Other forms of advice might be that there are no safety concerns at any intake that might occur through use as a supplement, because of the evidence of an absence of any toxicity even at levels of intake very much higher than would ever occur through dietary exposure of humans. The advice might be use-specific, that is, the substance might be considered to have no safety concerns in the context of the intake that is estimated from a certain specified use or uses. Should uses change, additional risk characterization might be necessary. A recommendation to reduce or avoid intakes by a certain sub-population (e.g. vitamin A supplements by pregnant women) might be necessary to ensure adequate health protection in such individuals. It might also be necessary to advise avoidance or modification of a certain production process or procedures (e.g. one method that was used to produce L-tryptophan).

The risk characterization should also include information on any factors that, exceptionally, might warrant exceedance of a health based guidance value.

Conclusions

There is a tendency to seek absolutes from risk assessment, that a substance is either safe or not. However, safety is not an absolute. It has been defined as the inverse of risk, and as risk is the probability of harm, it cannot be zero unless there is no exposure. Although deterministic approaches are often used in risk assessment, they are based on using default assumptions that are often conservative, and when several such assumptions are used, the overall assessment may be very conservative. However, in reality the inputs to the risk assessment are not individual values, but distributions and it is usually not possible to define the extreme tails with any accuracy. Hence, in theory at least, there could be a single individual who is extremely sensitive due to a unique combination of factors. Risk assessment has to strike a balance between what is desirable and what is reasonably achievable. This is particularly so in the case of dietary supplements when there needs to be some trade-off between the risk at high levels of intake and the benefits, or the risks of deficiency.

For some chemicals, for example drugs and pesticides, the risk assessor often has available a comprehensive dossier of information. However, for dietary supplements this is rarely the case, and it will be necessary to integrate information from a number of sources, which may include toxicological data from experimental studies in animals, information on adverse effects in humans, data on biomarkers from experimental animals or humans, and *in vitro* data and nutritional information from experimental animals or humans. The data package is unlikely to be complete, and the impact of the deficiencies on the risk assessment needs to be considered on a case-by-case basis.

Ongoing and future research on the development of informative, validated biomarkers of exposure, effect and susceptibility will help improve confidence in the risk assessments. These will enable more accurate information to be obtained in humans and will help bridge the gap between observations in experimental animals and humans. Similarly, greater insight into modes and mechanisms of toxicity will assist in inter-species extrapolation, identifying potential susceptible subgroups and in providing more informed advice on safe use patterns of supplements.

Risk assessment requires effective dialogue with risk managers. Problem formulation should be such that answers to the questions posed will provide answers that are useful to the risk manager in formulating options and advice to consumers. The output of the risk assessment needs to be in a form that is helpful to risk managers. It should be reasonably comprehensive, but not to the extent that the key information is submerged. The bases of the conclusions reached and the assumptions involved should be transparent, and the uncertainties in the assessment should be discussed and quantified to the extent possible. A statement of confidence in the risk assessment should be included, together with some indication of whether the estimates could be refined with additional effort or information.

References

Barlow SM, Greig JB, Bridges JW et al 2002 Hazard identification by methods of animal-based toxicology. Food Chem Toxicol 40:145–191

Barlow S, Renwick AG, Kleiner J et al 2006 Risk assessment of substances that are both genotoxic and carcinogenic report of an International Conference organized by EFSA and WHO with support of ILSI Europe. Food Chem Toxicol 44:1636–1650

Barton HA, Pastoor TP, Baetcke K et al 2006 The acquisition and application of absorption, distribution, metabolism, and excretion (ADME) data in agricultural chemical safety assessments. Crit Rev Toxicol 36:9–35

Boobis AR, Cohen SM, Dellarco V et al 2006 IPCS framework for analyzing the relevance of a cancer mode of action for humans. Crit Rev Toxicol 36:781–792

Cooper RL, Lamb JC, Barlow SM et al 2006 A tiered approach to life stages testing for agricultural chemical safety assessment. Crit Rev Toxicol 36:69–98

Dybing E, Doe J, Groten J et al 2002 Hazard characterisation of chemicals in food and diet. Dose response, mechanisms and extrapolation issues. Food Chem Toxicol 40:237–282

Edler L, Poirier K, Dourson M et al 2002 Mathematical modelling and quantitative methods. Food Chem Toxicol 40:283–326

EFSA 2006 Guidance of the Scientific Committee on a request from EFSA related to uncertainties in dietary exposure assessment. EFSA J 438:1–54

Eisenbrand G, Pool-Zobel B, Baker V et al 2002 Methods of in vitro toxicology. Food Chem Toxicol 40:193–236

European Parliament 2002 Directive 2002/46/EC of the European Parliament and of the Council. OJ L 183:51–57

Gibson RS, Perlas L, Hotz C 2006 Improving the bioavailability of nutrients in plant foods at the household level. Proc Nutr Soc 65:160–168

Gundert-Remy U, Dahl SG, Boobis A et al 2005 Molecular approaches to the identification of biomarkers of exposure and effect–report of an expert meeting organized by COST Action B15. Toxicol Lett 156:227–240

IPCS 2004 Principles for modelling dose-response for the risk assessment of chemicals (draft), WHO, Geneva http://www.who.int/ipcs/methods/harmonization/draft_document_for_comment.pdf

IPCS 2005 Chemical-specific adjustment factors for interspecies differences and human variability: Guidance document for use of data in dose/concentration response assessment, WHO, Geneva. http://whqlibdoc.who.int/publications/2005/9241546786_eng.pdf

IPCS 2006 A model for establishing upper levels of intake for nutrients and related substances, WHO, Geneva. http://www.who.int/ipcs/highlights/full_report.pdf

Kroes R, Muller D, Lambe J et al 2002 Assessment of intake from the diet. Food Chem Toxicol 40:327–385

Renwick AG, Barlow SM, Hertz-Picciotto I et al 2003 Risk characterisation of chemicals in food and diet. Food Chem Toxicol 41:1211–1271

Renwick AG, Flynn A, Fletcher RJ, Muller DJ, Tuijtelaars S, Verhagen H 2004 Risk-benefit analysis of micronutrients. Food Chem Toxicol 42:1903–1922

Said HM, Mohammed ZM 2006 Intestinal absorption of water-soluble vitamins: an update. Curr Opin Gastroenterol 22:140–146

Seed J, Carney EW, Corley RA et al 2005 Overview: Using mode of action and life stage information to evaluate the human relevance of animal toxicity data. Crit Rev Toxicol 35:664–672

Shitara Y, Horie T, Sugiyama Y 2006 Transporters as a determinant of drug clearance and tissue distribution. Eur J Pharm Sci 27:425–446

Singh SS 2006 Preclinical pharmacokinetics: an approach towards safer and efficacious drugs.
    Curr Drug Metab 7:165–182
Sonich-Mullin C, Fielder R, Wiltse J et al 2001 IPCS conceptual framework for evaluating a
    mode of action for chemical carcinogenesis. Regul Toxicol Pharmacol 34:146–152
van den Brandt P, Voorrips L, Hertz-Picciotto I et al 2002 The contribution of epidemiology.
    Food Chem Toxicol 40:387–424

DISCUSSION

Azzi: You mentioned the problem of the risk assessments for essential nutrients. We know of certain nutrients that vary in how essential they are for different functions. For example, a low dose of vitamin E (α-tocopherol) protects against the well known consequences of avitaminosis, but high doses may be important for Alzheimer's disease. So here there is a double risk assessment. There is another point I wanted to raise, synergism. You discussed a complex topic by taking one compound at a time, but sometimes these compounds have synergistic effects. For example, vitamin E and lycopene are synergistic.

Halliwell: I agree, but some nutrients can also be antagonistic, for example high-dose α-tocopherol and other tocopherols.

Boobis: I was using the term 'essential' in a fairly narrow sense. I was meaning 'to avoid nutrient deficiency'. The question I was really asking was, if you use the normal default on certain effects and divide by 100, for example the toxicological effects of vitamin E, does this take you below a level which is compatible with maintaining repleteness/sufficiency of vitamin E? In terms of the anti-Alzheimer's effect, the question is, do we specifically need vitamin E for that or can this by achieved by use of another substance? The toxicological concern is such that you could risk harm by giving a certain dose, even though it might also have a benefit. This is a different point from the essentiality of a mineral or vitamin that you cannot live without. Synergy is a vexed question. There are currently at least three international groups trying to get to grips with a framework that would allow us to address this issue. In the risk assessment community we don't have a good handle on how to address this. In terms of combinations, we need some way of deciding which combinations to start with, rather than looking at every possible combination.

Taylor: How would you see the model changing if you knew for sure that the substance you were studying was not essential? Is it the same kind of risk assessment for these types of substances? Related to this, what is 'adverse'? As we carried out the FAO/WHO workshop, one of the inputs we received early on was that we need to be careful in the world of nutrients about what adverse is. The question of reversibility and severity was something the workshop dealt with. The conclusion was that the point at which you provide public health protection may not be the point reflective of, for instance, the most severe outcome.

Boobis: For non-essentiality I think you can maintain the separation between risk management and benefit assessment. The question for risk assessment is what level do I believe is without harm? Then you go to benefit and ask will I get any benefit at that level? If not, you might want to go through another iteration and refine the risk assessment, or ask whether there is uncertainty about the toxicological hazard or exposure evaluation. But once you reach a firm conclusion the risk manager is stuck with that value, and then has to decide the trade-off between risk and benefit. Is the benefit I am going to get worth risking the associated degree of harm? It is the same with a therapeutic drug.

Taylor: Would you change your uncertainty factors if it was essential versus non-essential?

Boobis: Yes, that is what I was getting at. If something is genuinely essential such that its absence from the diet causes a deficiency syndrome, you have to trade off the toxicological properties and essentiality.

Taylor: The homeostatic mechanisms needed for essential nutrients would come into play there.

Boobis: In terms of adversity, that is a real issue. For marketed compounds, which are not given deliberately to people, the view is that almost any change is considered potentially adverse. I agree that we need a better definition of 'adverse'. The IPCS (International Programme on Chemical Safety) has a definition that is along the lines that it impairs the capacity to thrive, survive and reproduce, and reversibility doesn't come into it. If you get a headache, it is an adverse effect even though it is completely reversible. So we need a more subtle definition that doesn't just use irreversibility as the catch-all.

Katan: Does toxicological assessment meet the criteria for evidence-based medicine? Does it predict harm or actual disease in humans? My impression is that the sensitivity and specificity of toxicological classifications for predicting actual human disease is low. Can we improve this process? One thing I have been thinking about is that pharmaceutical companies probably have lots of data on substances which they have first tested toxicologically in animals and then tested in humans. The problem is, many of these data will be confidential.

Boobis: I know of a study that is underway that is trying to address this question. They asked all the pharmaceutical companies involved in the consortium (six or eight) to put into the database all the preclinical toxicology information about every compound in development over a specific period (some of these drugs would proceed to trials, some would make it to the market). They then looked at what was observed in people and compared the two, to determine the predictability of the animal toxicology to humans. In terms of toxicological assessment being evidence-based, it is, in that we look for effects that we determine experimentally. But you are right that we are making assumptions about the relevance to humans. The default assumption is that in the absence of information to the contrary, the effect is relevant. If we see an effect in a rat or a dog, we assume that this effect would occur in people, usually allowing for 100-fold greater sensitivity of a sensitive subject than the experimental animal. It is a conservative assumption, based on a default position that in the absence of information to the contrary we will assume that this should certainly be protective.

Katan: It could be the other way round, of course. Humans could be 100 times less sensitive.

Boobis: Absolutely. This is what the supplements people say to us. Effects seen in a rat won't necessarily be seen in humans.

Halliwell: I think you have made some fundamental points. What you have is a theoretical framework. We should address this again as we come to each individual nutrient.

Yetley: You presented a graph showing a curve for the risk of deficiency and a separate curve for the risk of cancer for a nutrient. The way you drew your graph is that the two curves are overlapping. The way that the curves are usually drawn in a nutritional setting is that there is a gap in between them to represent homeostatic mechanisms. Is this just semantics, or is there a different perspective between nutritionists and toxicologists?

Boobis: It will vary with the substance. For some there will be a substantial separation of the curves, and for others the curves will be close together or even overlap. With vitamin A in some populations, you probably can't separate those curves very well.

Yetley: Is it appropriate to assume that they overlap?

Boobis: We don't need to make any assumptions, because we can define the curves. Where they fall may leave a gap, or there may be an overlap. The problem is that these are Gaussian (normal) probability curves, and in theory they have no intercept on the x axis. This assumption would imply that there is no such thing as zero risk.

Aggett: An important issue in understanding these curves is the event or criteria one plots them against. If one is taking serious evidence of deficiency and serious evidence of hazard, there will be a plateau in between.

Azzi: The deficiency curve should actually be substituted by multiple curves. In fact, the consequences of deficiency for different endpoints need each their separate curve. Hazard could also be represented by different curves. For example, if a patient has cancer and takes adriamycin, is the loss of hair a big hazard with respect to the cure of cancer?

Boobis: I am not a nutritionist. As a toxicologist, I only have one pivotal hazard curve. We take a critical endpoint that we think is relevant and most sensitive, and we do a risk assessment using that curve.

Taylor: But it would be possible to have many curves for hazards if you were picking different critical endpoints. Would you never consider looking at multiple curves for critical endpoints?

Boobis: The risk manager would have to give specific guidance as to the question being asked. If you are asking 'what is the risk?' then the risk assessor would look at all the data and come to a conclusion on the critical endpoint. If you say, 'I want you to look at severity of risk as a factor and a specific subpopulation risk', you might tailor the risk assessment to different outputs, resulting in a family of curves. A good example is fluoride. Dental fluorosis is regarded by many to be just a cosmetic effect, not an adverse one. However, skeletal fluorosis is an adverse effect. Effects at higher levels are even more adverse. You could have a family of curves and the risk manager would then decide where they wanted to place fluoride.

Aggett: That's a good point. The IPCS assessment of fluoride experienced this problem. Dental fluorosis appears to precede skeletal but trying to understand if this is true so that the mottling of teeth could be placed in a sequence of adverse events and, possibly explored as a predictive marker of skeletal fluorosis was not possible (this was also difficult because there was no good means of knowing levels of intake nor the duration of this exposure). Do you conceive that it may be possible to redefine critical events in risk assessment so that one can balance responses to low intakes and reactions to high intakes on the bases of homeostatic metabolic reactions, rather than on the longer-term and more extreme outcomes?

Boobis: I think that we should consider all endpoints relevant to the duration of exposure of concern. There is a move, with the development of biomarkers, to try to detect changes that, whilst not themselves adverse, may be precursors of adverse effects. This will require some consideration of the appropriate model (uncertainty factors) for extrapolation to humans.

Halliwell: One thing that has always worried me about nutrients, is that people who take supplements are not 'normal'. They are the top end of the population: they are generally well nourished and health conscious. This is different from fortification of a staple nutrient where everyone is getting it.

Setting dietary intake levels: problems and pitfalls

Robert M. Russell

Jean Mayer USDA Human Nutrition Research Center on Aging, School of Medicine and Friedman School of Nutrition Science and Policy, Tufts University, 711 Washington Street, Boston, MA 02111-1524, USA

Abstract. Recommended dietary intake levels are the nutrient standards used in designing food assistance programmes, institutional feeding programmes, counselling and teaching. In the USA, the recommended dietary allowances (RDAs) are the basis for setting the poverty threshold and food stamp allotments. In the 1990s, a new paradigm was put forth for estimating nutrient requirements and recommended intake levels. This considered the level of nutrient needed for normal body functioning (versus the amount needed to prevent a deficiency state from occurring). An estimated average requirement (EAR), an RDA and a tolerable upper intake level (UL) were determined for most nutrients. In setting forth these nutrient intake levels (dietary reference intakes, DRIs), a number of data challenges were encountered. For example, it was recognized that for most nutrients there was an absence of dose–response data, and few chronic human or animal studies had been undertaken. In considering how to revise nutrient intake recommendations for populations in the future, the following pitfalls must be overcome: (1) invalid assumption that a threshold level for a requirement will hold for all nutrients; (2) lack of uniform criteria for the selection of the endpoints used (need for evidence-based review, consideration of comparative risk); (3) invalid extrapolations to children for many nutrients; (4) lack of information on variability of responses, and interactions with other nutrients; and (5) lack of understanding in the community of how to use the various DRI numbers.

2007 Dietary supplements and health. Wiley, Chichester (Novartis Foundation Symposium 282) p 29–45

The scope of this paper is to describe how the recent US dietary intake levels (dietary reference intakes, DRIs) were derived and what challenges, problems and pitfalls arose during the process of setting nutrient intake goals for individuals as well as for the planning of public health programs, such as food assistance and educational programs. It is also important to remember that dietary intake goals are needed for making policy decisions regarding food fortification, supplementation, agriculture and food labelling.

29

The recommended dietary allowances (RDAs) in the USA were first established in the 1940s in order to provide adequate nutrition for the military as well as for the civilian population during periods of war and/or economic depression. It had been realized that many potential military recruits were being turned down by the army because of poor nutrition and physical fitness. However, it was envisioned that the RDAs would be used not only by the military but also as standards for civilian public health programmes. In 1941, the first RDAs were determined in the USA for energy, protein, calcium, iron, and the vitamins A, C, D, E, thiamine, riboflavin and niacin (National Research Council 1943). The science base for establishing the original RDAs consisted of observations of usual food patterns in apparently healthy populations as well as some experimentally derived nutrient requirements (e.g. balance studies). Subsequent revisions of the RDAs included more nutrients, but essentially used the same methodologies in deriving the numbers. However, by the 1990s, some new thinking began on nutrient standards with regard to the following:

(1) A need to consider chronic disease endpoints when establishing nutrient requirements (for example, how much of a nutrient does it take to prevent a chronic disease rather than to prevent a deficiency state from occurring).
(2) A need for greater clarity in selecting indicators of adequacy (endpoints) for a given nutrient.
(3) The need to consider other nutritional components such as fibre, in addition to the usual macronutrients and micronutrients.
(4) The need to delineate the appropriate uses of the derived nutrient standard recommendations.
(5) The need to apply risk assessment to nutrients (i.e. the establishment of upper levels) in an era of enthusiastic supplement use.

In 1994, a concept paper was published by the Institute of Medicine of the National Academy of Science of the USA to establish a framework to develop nutrient intake recommendations that would meet a variety of uses, and to base nutrient requirements on the reduction of chronic disease risk with a clear rationale for the endpoints chosen (Food and Nutrition Board 1994). In addition, the format that was proposed included a review of other food components, such as fibre, and provided for estimates of tolerable upper levels of intakes. Thus it was envisioned that the DRIs would include several reference numbers, not just an RDA.

The process of setting the DRIs included convening a panel of experts for a particular set of nutrients in order to review the evidence, not only by a rigorous review of the literature, but also by the holding of workshops at which other experts in the field could speak, including workshops held at scientific meetings such as at the Experimental Biology meetings that occur in the spring of each year. Once the report of the expert panel was established, the report then underwent rigorous external review. As mentioned above, the DRI framework was to include multiple reference points, including an estimated average requirement (EAR), an RDA, and a tolerable upper intake level (UL) (Fig. 1). In situations where there were inadequate data to establish an EAR, an adequate intake level (AI) was determined, a much more vague and not very useful number in terms of either an individual goal or for public health planning. The DRI panels that were convened started with the panel on bone-relevant nutrients (calcium, vitamin D, phosphorus, magnesium and fluoride). This was followed by the panel on B vitamins and choline; a panel on antioxidants (vitamins C and E, selenium, β carotene and other carotenoids); a panel on vitamins A and K and micronutrients (arsenic, boron, chromium, copper, iron, iodine, manganese, molybdenum, nickel, silicon, vanadium and zinc); a panel on macronutrients; and finally, a panel on electrolytes and water. In addition, other panels were convened including an 'Upper Reference Levels Subcommittee' and a 'Uses of DRIs Subcommittee' for both assessment and planning purposes.

   The scientific databases available for setting dietary reference intake levels are clinical trials, epidemiological observations, balance studies, depletion/repletion studies, animal experiments, biochemical measurement, and observed intakes in healthy populations. The DRI concept is illustrated in Fig. 1. The EAR is the observed level of intake at which, for the given indicator of adequacy (e.g. bone mineral density), half the population would be considered deficient and half would be considered sufficient for a given nutrient (e.g. calcium, vitamin D). The RDA was then set at a level considered to be two standard deviations above the EAR,

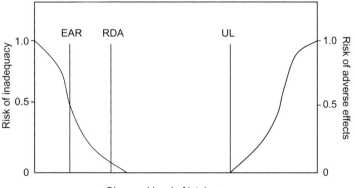

FIG. 1.   Dietary reference intakes: applications in dietary assessment. Reprinted with permission from Food and Nutrition Board (2000) © the National Academy of Sciences, Courtesy of the National Academies Press, Washington, DC.

so that a person eating this intake level would be very unlikely to be deficient. The panels were also asked to look at establishing EARs for different age groups and sexes. The upper level was set at the highest amount of nutrient intake at which no toxicity would be expected. However, going above this intake level on a daily basis would increase the risk of an adverse effect.

The EAR is the key number to establish from a public health point of view (and even possibly from an individual's point of view). Whereas it was hoped that the EARs would be based on the requirements for prevention of specific chronic diseases (i.e. a disease endpoint), only a few EARs were actually established on this basis. Several EARs were established using specific biochemical functions as endpoints, or less specific physiological outcomes (e.g. red cell haemolysis for vitamin E). Whenever a functional criterion was used, it was envisioned that it would be associated with some health benefit.

Using vitamin A as an example, the various indicators that were considered for establishing an EAR for this vitamin were dark adaptation, plasma retinol concentration, isotope dilution experiments for body stores, relative dose–response/modified dose–response tests, conjunctival impression cytology, and immune function (Food and Nutrition Board 2001). However, for the most part, dose–response data were not available on these parameters, with the exception of dark adaptation. In considering dark adaptation, four studies were evaluated using various doses of vitamin A in 13 individuals. Using these four studies, it was determined that 50% of people would be abnormal and 50% would be normal with regard to dark adaptation at a vitamin A (retinol) intake level of 300 µg per day. This, then, would be the EAR. However, the coefficient of variation was extremely large for this indicator (40%), and therefore, an RDA was not established using dark adaptation. The panel therefore chose a different approach to establish an EAR for vitamin A: a factorial approach. The factorial approach considered the mean percent of vitamin A stores lost per day, the minimal acceptable liver reserve (an amount to provide 3–4 months of vitamin A to an individual while eating a diet lacking in vitamin A), the weight of the individual, the ratio of total body to liver vitamin A reserves, and the efficiency of storage of ingested vitamin A. This approach yielded an EAR of 625 µg per day for an adult male greater than 18 years old (more than twice the EAR determined by using dark adaptation as the indicator). For females, because of the lower reference weight, the EAR was established at 500 µg per day. Using a coefficient variation of 20% (based on the variance of the half life of vitamin A), the RDA for adult males was set at 900 µg per day, and for adult females at 700 µg per day. The EARs and RDAs for younger age groups (greater than one year old) were based on extrapolation formulas from adults, whereas for babies under 12 months of age, an adequate intake level was established based on human milk content for 0–6 months old and extrapolated upwards for children 7–12 months old from the 0–6 month data.

Continuing with using vitamin A as an example, let's take a look at how the tolerable upper intake level was derived. The process of determining a tolerable upper level is: (1) identify the hazard of taking too much of a given nutrient (e.g. liver enzyme elevations), (2) perform a dose–response assessment to include the identification of either a no-adverse effect level (the highest level at which no adverse effect is seen = NOAEL) or lowest-adverse effect level (the lowest dose at which an adverse effect is seen = LOAEL), (3) perform an uncertainty assessment, (4) perform an exposure assessment for the population of interest, and, finally, determine a risk characterization. The upper level is established by dividing either the LOAEL or NOAEL by an uncertainty factor. The basis of the uncertainty factor is quite subjective, but the panel setting the upper levels for each nutrient had to consider interindividual variation (the source of uncertainty would be greater with large variation in response); animal to human extrapolations (the uncertainty factor would be a greater number if based on animal experiments); short-term versus chronic exposures (the uncertainty would be greater for short term exposure); use of a LOAEL instead of a NOAEL (the uncertainly would be greater for a LOAEL as it is considered a less 'safe' target than an NOAEL); the number of subjects studied (if only a small number of subject were studied, the uncertainty would be greater than when a large number of subjects were studied); and finally, the severity of the effect [the uncertainty factor would be set high if the hazard is of a serious nature (e.g. liver failure) versus of a trivial nature (e.g. flushing)]. For vitamin A, the adverse effects that were considered in setting the upper level were bone mineral density, liver toxicity, teratogenicity (for women of reproductive age), and bulging fontanels in infants. For women of reproductive age, teratogenicity was chosen as the hazard, and a NOAEL was established at 4500 µg per day. A very large number of subjects had been studied with regard to teratogenicity (Rothman et al 1995) so the uncertainty factor was set relatively low at 1.5, so that the final upper level was set at 4500 ÷ 1.5, which equals 3000 µg per day. For other adults, liver toxicity was used as the hazard, and the LOAEL was set at 14000 µg per day (Minuk et al 1988, Zafrani et al 1984). Because only a small number of subjects had been studied with large individual variation and because the severity of the effect was quite profound, the uncertainly factor was set at five. Therefore, the UL was 14000 ÷ 5 ≈ 3000 µg per day. Bone mineral density was not chosen as the hazard for taking too much vitamin A because of the conflicting data in literature at the time when the panel was considering the vitamin A upper level.

As seen from the example of vitamin A, there are many challenges in setting DRIs using chronic disease indicators. First of all, there is often a lack of specificity of the indicators and therefore doubts about a purported nutrient causing an effect on the indicator. For example, it was known that there are data to implicate abnormal immune function as an early indicator of vitamin A deficiency in deprived populations (Food and Nutrition Board 2001). However, this indicator is influenced by so many factors (e.g. other nutrient deficiencies, stress, etc.), that it could not be certain that the effect on immune function was solely due to vitamin A deficiency. Furthermore, there were no human studies using controlled diets containing different amounts of vitamin A with immune function being used as the indicator for adequacy. In fact, for *most* physiological and/or chronic disease endpoints there is a lack of data on the responses to multiple levels of the nutrient in the same individuals; that is, there is a lack of dose–response data. Therefore, an EAR was often impossible to establish using such endpoints, and only an AI could therefore be derived. Thirdly, it seems for most nutrients there is a lack of information on variability of the response on the indicator. Thus, in most cases a coefficient variation of 10% was assumed, for which there was not a good basis. Fourthly, there was lack of information in most experiments on interactions with other nutrients. Fifth, there was a lack of information on children, adolescents and the elderly for which to establish EARs with any degree of certainty. Sixth, there were few available *chronic* human or animal studies. Most of the studies were acute, lasting one or two weeks. Seventh, there were few surveillance studies to establish a NOAEL. Eighth, many databases ignored supplemental nutrient intakes and looked at dietary nutrient levels from food only. Finally, there was no consideration in most studies with regard to differences in bioavailability of the nutrients depending on the food matrix.

From the US experience, the main pitfalls in setting dietary intake levels are as follows:

(1) The threshold level for requirement is not necessarily valid for all nutrients (e.g. as is the case for carbohydrate, fat and protein).
(2) There was lack of uniform criteria for selection of the endpoints used for judging adequacy or toxicity; some of the endpoints chosen were disease endpoints, some were biochemical and some were factorial.
(3) The extrapolations to children were often not valid, but were used due to the fact that there is such a dearth of studies on children.
(4) The DRI numbers (EARs, RDAs, and AIs) and their uses are not well understood and are debated by the experts even now.

An example of a specific issue that has been encountered with regard to the DRIs in the US, is that for several nutrients (e.g. vitamins D, E and potassium) the majority of the population is not coming anywhere near meeting the RDA (Food and Nutrition Board 1997, 2000). Does this mean that the population is unhealthy or malnourished with regard to these nutrients, as seems to be the case with vitamin D, or that the RDA was set too high? Another example of an issue is that the upper level for vitamin A set for infants has been questioned due the fact that 56% of formula-fed infants age 4–5 months old in the WIC program are eating above the established UL (Food and Nutrition Board

2006). Does this mean that the UL has been set too low or does this mean that the amount of preformed vitamin A in infant formulas is too high?

Finally, there is the issue that science has moved on since the establishment of these latest DRI values. How should new DRIs be derived using our past experience in establishing the older values? There is new information on many nutrients which could alter the recommendations made just 10 years ago. The best example of this is vitamin D where several lines of data point to an even higher requirement than was set by the DRI committee in 1998 (Heaney et al 2003, Glerup et al 2000, Food and Nutrition Board 2005, Weaver & Fleet 2005). It has not been established exactly what the process is going to be for revising DRI numbers in the USA, as the process is extremely costly and time consuming. It is hoped that when there is a predominance of new data of public health importance that would affect the RDA level, a committee can be convened to reconsider the DRIs for that particular nutrient. However, the mechanism for doing this is not in place.

As stated before, the DRIs are important numbers to be used as nutrient goals for individuals, for doing assessments of dietary accuracy, for planning and procuring food supplies, for the establishment of food fortification and supplementation policies, for the planning of education programmes, for the planning and evaluation of food assistance programs, for food labelling policies, for agricultural policies, and finally, for dietary guidance policies (dietary guidelines). We have moved forward a great deal from the era when the RDA was the only published number. It is now becoming recognized that the EAR is probably the key number to consider, for populations and possibly even for individuals (since the coefficient of variation around the EAR, for most nutrients, is assumed and is not known). Therefore, the RDA number may be rather artificial. The framework used in the 1995–2004 US revisions of the DRIs needs to be carefully examined as we now approach the need to change the values for certain nutrients (e.g. vitamin D). The documentation and transparency for choices made by the expert panels (e.g. on the choice of an indicator of adequacy) needs to be clarified, and uniform methodology needs to be established for arriving at a choice. Finally, the uses to which the DRI numbers have been put and the problems that arise when trying to apply them needs close examination. We have to recognize these challenges and fill in many data gaps (e.g. dietary requirements for children, adolescents, and elderly) before we can achieve truly optimal health by dietary means.

References

Food and Nutrition Board 1994 How should the RDAs be revised? National Academy Press, Washington DC

Food and Nutrition Board 1997 Dietary reference intakes for calcium, phosphorus, magnesium, Vitamin D, and fluoride. Food and Nutrition Board, Institute of Medicine, National Academy Press, Washington DC

Food and Nutrition Board 2000 Dietary reference intakes for Vitamin C, Vitamin E, Selenium, and carotenoids. Food and Nutrition Board, Institute of Medicine, National Academy Press, Washington DC

Food and Nutrition Board 2001 Dietary reference intakes for Vitamin A, Vitamin K, arsenic, boron, chromium, copper, iodine, iron, manganese, molybdenum, nickel, silicon, vanadium, and zinc. Food and Nutrition Board, Institute of Medicine, National Academy Press, Washington DC

Food and Nutrition Board 2005 Dietary reference intakes for water, potassium, sodium, chloride, and sulfate. Food and Nutrition Board, Institute of Medicine, National Academies Press, Washington DC

Food and Nutrition Board 2006 WIC Food packages: Time for a change. Institute of Medicine of the National Academies, National Academies Press, Washington DC

Glerup H, Mikkelsen K, Poulsen L et al 2000 Commonly recommended daily intake of vitamin D is not sufficient if sunlight exposure is limited. J Internal Med 247:260–268

Heaney RP, Davies KM, Chen TC, Holick MF, Barger-Lux MJ 2003 Human serum 25-hydroxycholecalciferol response to extended oral dosing with cholecalciferol. Am J Clin Nutr 77:204–210

Minuk GY, Kelly JK, Hwang WS 1988 Vitamin A hepatotoxicity in multiple family members. Hepatology 8:272–275

National Research Council 1943 Recommended dietary allowances. Washington, DC: National Research Council

Rothman KJ, Moore LL, Singer MR, Nguygen UDT, Mannino S, Milunsky B 1995 Teratogenicity of high vitamin A intake. N Engl J Med 333:1369–1373

US Department of Health and Human Services and US Department of Agriculture 2005 Dietary Guidelines for Americans. Available at: *http://www.health.gov/dietaryguidelines/dga2005/document/pdf/DGA2005.pdf*

Weaver CM, Fleet JC 2005 Vitamin D requirements: current and future. Am J Clin Nutr 2004 80(Suppl):1735S–1739 (erratum: 2005 Am J Clin Nutr 81:729)

Zafrani ES, Bernuau D, Feldmann G 1984 Peliosis-like ultrastructural changes of the hepatic sinusoids in human chronic hypervitaminosis A: report of three cases. Hum Pathol 15:1166–1170

DISCUSSION

Azzi: One nutrient that is part of almost everyone's diet is alcohol. What is being done in terms of giving guidelines for intake? Some people suggest there are health benefits, while others say it is always toxic.

Russell: This is one nutrient that the Institute of Medicine has shied away from. There is little funding available for studying the benefits of alcohol. The guideline given is that if you must drink, then do it moderately. But this guideline is vague and there isn't much science behind it.

Scott: One of the concerns is the acceleration or development of tumours. The natural history of colon cancer, for example, takes place over many years. These data aren't used in the studies looking at acute effects. What comfort is there in the fact that a significant proportion of adults in the USA over the last three decades have used supplements, so there has been quite a dramatic difference in exposure? Colon cancer is likely to be a multifactorial disease, but on the other hand, if one third of people are using supplements containing 400 µg/day of folic acid and have been doing this for 30 years, is it reasonable to say that any difference might have emerged?

Russell: With regard to folate, such data were not considered in coming up with an upper limit. For someone with hidden cancer, we wouldn't be able to come up with a recommendation unless it were to be done on the basis of age (cancer being an age-related disease). The issue is how do you identify a substantial group who could potentially be harmed by a level of a nutrient that others might benefit from?

Scott: The upper limit for folic acid was based on folic acid masking or precipitating pernicious anaemia. It is probably a vanishingly rare event. The real issue relates to the bigger risk of the development of cancer. Is it true that there are good data that adult Americans for the last two or three decades have taken supplements containing 400 µg folic acid?

Russell: Yes.

Scott: I get some comfort from that.

Yetley: It's true that Americans have taken folic acid, but the use pattern has changed. Unlike any other nutrient, folic acid has been regulated in the USA as a food additive, so there has never been more than 400 µg added to a baseline diet. I appreciated you saying that you have no choice, but you have to set a DRI because there are many applications. Yet in many cases you are dealing with a paucity of evidence. Folic acid is a good example. Before we did fortification, we considered a lot of factors such as increased risk of cancer and interactions with anti-folate drugs. We had virtually no data on safety. The only data, even though they were scarce, were on B12 absorption, so this became our endpoint. It is a trivial endpoint, but it is useful because it is the only data we have. If we look at β carotene before the lung cancer study, there would have been a trivial endpoint for ULs for β carotene. What bothers me is that when people claim we have a trivial endpoint for the UL and a more serious endpoint for the RDA, so we have to balance that. I think we also have to look at whether we have a trivial endpoint for the UL because we have no data, or do we have enough data to make a wise judgement? This imbalance of data is troublesome.

Halliwell: You said that for most of the vitamins and minerals we don't have an adequate database to set an RDA or UL. For which ones do we have an adequate database?

Rayman: It was thought we had a reasonably adequate database for selenium, in so far as we were looking at optimisation of glutathione peroxidase activity. Now we are having to revisit this, because we realize that this isn't the right criterion to use. There are other selenoproteins that have higher requirements than glutathione peroxidase. But this was one nutrient where there seemed to be quite a good biochemical or metabolic endpoint.

Boobis: I'm interested in the lack of a threshold for cholesterol that you mentioned. How secure is this observation? Cholesterol is an essential compound in the body, and low levels of cholesterol intake are associated with risks as well.

Russell: The body synthesizes most of the cholesterol that is needed. The panel felt that it couldn't determine an upper level threshold for cholesterol because there is a direct relationship between risk and LDL cholesterol levels in the blood, which increase with cholesterol intake. The same is true for Na^+, which has a direct relationship to blood pressure. It is hard to pick a threshold where you can say these nutrients are safe.

Katan: I don't think that by itself eliminating cholesterol from the diet would cause harm. Speculations that low serum cholesterol concentrations or low dietary cholesterol intakes could be harmful have been largely refuted (Steinberg 2006). However, in practice you cannot eliminate cholesterol from the diet without eliminating all meat, fish, milk, cheese and eggs, which causes other problems.

Rayman: Total cholesterol levels aren't a terribly good discriminator of cardiovascular risk. You need to look at LDL subfractions.

Katan: That is quite a different issue from cholesterol in foods. Dietary cholesterol only has a small effect on the levels in the blood.

Cashman: Considering the first round of DRIs for Ca^{2+} and vitamin D are almost 10 years old, do you know when these will be revised?

Russell: First we are trying to determine what changes we need to make to the DRI framework and model. Vitamin D and Ca^{2+} are the first nutrients on the radar screen because there is now a lot of additional evidence. The current DRIs for vitamin D are probably incorrect, i.e. too low. The IOM will likely review these in the next couple of years.

Cashman: In relation to vitamin D recommendations, it would be ideal to get input from dermatologists and skin biologists. We need to revisit the relevance of sun exposure versus diet.

Russell: That will be a big debate. In Boston an endocrinologist got his appointment taken away for advocating sun exposure. There is a common sense approach, which involves a small amount of sun exposure per day. There is no evidence that this will increase melanoma risk to my knowledge.

Cashman: It is probably important to get nutritionists and dermatologists to discuss this because this could lead to conflicting public health messages.

Russell: We try to include as many voices as possible in these processes.

Taylor: I have a comment about the AI versus the RDA. In some meetings people have been suggesting that RDA is perhaps unnecessary, and all that is needed is the AI. As I listen to this group I'm thinking how important the RDA is as we start to think about upper levels.

Russell: Wherever there is an RDA established, if you look at what the population is eating, which would be an AI, it is almost always above the RDA, at least in the USA. If you are eating below the AI you have no idea of where you are on the spectrum of things, because you don't have an EAR.

Taylor: But what if you flip it and talk about the EAR being the only thing you set, and then moving from the EAR to an RDA. There are people asking why we are doing RDAs when we already have an EAR.

Russell: In a layperson's point of view it might be a bit odd to set the EAR as the goal. At that goal, 50% of the population will be eating inadequately. This is why I think it would be a hard number to use. How could one explain to the public the justification for using the EAR as the goal?

Taylor: But then the argument is that 2.5 SDs above that is an arbitrary figure.

Russell: Yes, it is, since we don't know what the coefficient of variation is around the EAR for most nutrients.

Katan: I share your concern about the scientific basis of the DRIs. But before the message gets out to the public that all these numbers are without a scientific basis, what would the scenario be if they were abolished? What would the effect on the health of populations be if we said that we don't know anything?

Russell: I think it would be a disaster.

Shekelle: Why would it be a disaster? Most people are eating as much as they need of these things.

Russell: In the USA there is a lot of interest in nutrition. People want to have enough vitamins and minerals. If you suddenly tell people that all the information is bogus, it will add to the feeling that nutritionists don't know what they are doing. They would lose confidence. If they just eat whatever they want, who knows what the long term health consequences would be? These are numbers the public wants: they ask for them.

Yetley: DRIs can be important in policy decisions. For example, DRIs become the basis for setting nutritional standards for school lunch programmes. If you didn't have an RDA for vitamin A for school age children, for example, school lunch standards may not be able to require that school lunches contain a specified amount of vitamin A. The absence of a RDA for vitamin A for this age group would likely mean that this nutrient is no longer considered in this type of policy application. Therefore, not having a RDA or UL for a known nutrient could potentially be a more serious public health risk than a RDA or UL developed by a well documented process whereby qualified experts use their scientific expertise and judgment to develop nutrient reference values based on evidence that is deemed sufficient for this task but may be less than ideal. For essential nutrients, it is important to note that the uncertainty is not about the essentiality of these nutrients for optimal health; this is well documented and accepted by the scientific community. Rather, the uncertainty relates to the best endpoint and its associated intake level to use as the basis for setting a reference value. Additionally, it is often necessary to extrapolate results from studied to unstudied groups (e.g. from young adults to school age children) because primary studies are often not available for all the age/gender/lifestage (e.g. pregnancy) groups of interest. This also can lead to uncertainties.

Ernst: Would it be worse not to have any DRIs versus DRIs that are demonstrably wrong?

Russell: I am saying the scientific basis for many of the DRIs is relatively weak; not that it is wrong. We should be giving the public and health professionals some notion of what our confidence in the data is. I believe we should have some kind of weighting behind the numbers, that practitioners can get a feel for how many data lie behind each number.

Ernst: What was impressive was how convincingly you showed the lack of data. Whenever one is faced with a lack of data, there is only one way of proceeding: that is, to fill the gaps. I wonder whether the IOM is using its quite considerable influence to fill these gaps. Were these original DRIs characterized as being 'guesstimates', which they seem to be? Is there a proper schedule for doing the science and reassessing them?

Przyrembel: The AIs are dynamic numbers. Because of the increasing fortification of foods and of the increasing numbers of the population taking supplements, the AI has definitely increased. On the other hand, your data about the zinc intake of the formula-fed infants do not prove that the UL is wrong. Perhaps it is not normal to feed infants with formula, and perhaps one should take the breast-fed infant as the norm in this case. There is a difference between the bioavailability of zinc from breast milk and formula. This is not a proof that the UL is wrong. It just proves that with such an intake, a formula-fed infant does not show adverse effects.

Azzi: The change in lifestyles is important. On one hand we have rising obesity, and on the other we have increasing numbers of marathon runners. How does this affect recommendations?

Russell: With regard to life activity, the nutrients most affected are the macronutrients. There are tables for judging this in the RDA report on macronutrients; that is, people can match up their activity level with what they should be consuming. The panels looking at micronutrients concluded that there is no evidence that micronutrients needs are affected by activity levels.

Azzi: What is being done with vitamin K, and the assumption that vitamin K has an important role in the prevention of osteoporosis?

Russell: At the time the panel met there was not enough evidence with regard to bone effects from vitamin K. This is an evolving story, but we have no real intervention studies on this yet, just epidemiological relational studies.

Coates: I wanted to take a bit of the heat off the IOM. Edzard Ernst, your comment suggested that the IOM had a major responsibility for resetting these levels. There is no question that there is a major role for them, but one of the things the IOM panels did in constructing the DRIs last time (which will be informative in developing future iterations of this) is that at the end of each of the chapters they put in a range of research recommendations identifying major gaps of subpopulations where more data needed to be gathered. It is the responsibility of the entire community to take a fresh look at this. We have commissioned the IOM to gather these research recommendations into an accessible database that others can use. One of your last messages was that, in general, it has been moving well. We are recognizing the responsibility our federal government has in stimulating those kinds of activities. Frankly, there are some issues which could be improved, but we are trying to address these.

Halliwell: People will only be able to do the research if someone pays for it.

Coates: They will also only be able to redo these DRIs if someone pays. It does come back to some common themes.

Wharton: You said once or twice that the AIs are somewhat above the RDAs. If we look at the data worldwide for Ca^{2+}, the actual observed intakes are way under the RDA in a considerable proportion of the population. With the exception that Ca^{2+} may play a role in some occurences of rickets in the sub-Sahara (e.g. Thacher et al 1999), I am not aware that there are vast numbers of people in developing countries with metabolic bone disease. Are these sorts of data taken into account when you consider a nutrient?

Russell: In the USA, only AIs were established for Ca^{2+} and vitamin D—not EARs or RDAs. To clarify, the EARs and RDAs were specifically designed for the USA and Canada. It is clearly stated that this doesn't apply to developing countries. The only area I am aware of where a developing country number was talked about was for vitamin A. The EAR was placed at around 300 µg/day, which is half of what the EAR is for adults in the USA.

Wharton: How would you respond to the argument that these people with intakes of Ca^{2+} considerably below the DRIs seem to be disease free?

Russell: This could have to do with the whole diet. There will be a complex mix of different nutrients and substances (such as fibre) in a diet, which could result in a whole raft of bioavailability differences. DRI numbers have to be region specific.

Halliwell: We could tell it was a US population when your slide had an average body mass of 76 kg!

Aggett: I agree that the values that the panels have come up with are comfortably in the right ball park, but there are some interesting challenges. Selenium is one that has been mentioned. Ca^{2+} is another. One of the significant things that emerged from the UK dietary reference values, and from the EU population reference intakes, was that the suggested calcium values were considerably less than those of North America. However, the recommendations in Europe coincided well with observed intakes. But, in fact, some population groups exceed observed intakes. The reason why these more conservative values emerged in the UK and Europe is that we considered issues relating to bioavailability of Ca^{2+}, and we were informed by the experience of people working in regions where incident sunlight was quite high.

Coming back to the issue of iron and availability, this is an intriguing problem when it comes to infant formulas. Historically, there is a large amount of zinc in infant formulas, and this was introduced when they also contained a large amount of iron. The amount of iron in an infant formula was considerably more than there was in maternal breast milk. The zinc:iron molar ratio in human breast milk is about 3:1. In the 1960s and 1970s in infant formulas the ratio was reversed: there was 6:1 iron:zinc as a molar ratio. Since we are aware of the importance of zinc to growth, it is interesting that some infants were found to have improved growth when given zinc supplements. One can hypothesize that what was happening was the amount of iron in those formulas was interfering with absorption and utilization of the zinc that was present. The reason the iron was there was because negative chemical balance studies done in infants in the 1930s created concern about potential iron deficiency in infants. However, we know now that infants can absorb iron if they need it, and most of the negative iron imbalance was a manifestation of intestinal homeostasis. One can take the historical precedent for setting some micronutrient requirements, but at the same time we must take the total current physiological understanding into consideration as well.

Scott: I want to comment on the lack of evidence for setting the ULs. The suggestion was made that it was up to the scientific community to come up with experimentation to create more data. Of course, the problems are obvious: there will be ethical considerations; they are not just questions of funding. But the natural experiment of people who take supplements against people who don't, in very large numbers, would give us significant differences in most nutrients. Are there surveillance programs in place to look for the emergence of disease patterns that might indicate risk? This is a big natural experiment that we know is taking place.

Russell: The Health and Nutrition Examination Survey (HANES), which measures nutrient status/intake and health status in a cross-section of the population, comes closest to this. It looks for disease trends (US Department of Health & Human Services 1996).

Scott: There's some comfort if they don't see them.

Taylor: This assumes you are measuring carefully what these people were taking. One of the problems that exists is that we are not collecting the right kind of data on the intake. Perhaps you are collecting end-points, but you don't really know whether each person took one tablet a day, or whatever. I am not sure the database is anywhere near the point it can be used to address this question. I want to go back to the issue about the public needing to know the state of the science around these values. I go back to the qualified health claims experience, where the claims are ranked in terms of supporting science and the consumer is told the quality of the science. This has the potential to be problematic. What is the consumer to do with advice on intake that says we think you should take this much selenium, but we don't know how right that number is? Can it give the consumer a confused message? It also plays into the idea that the science changes every two minutes and therefore cannot be trusted. Is it better to think in terms of a public health protection policy that filters this out, so the consumer is not being confused.

Halliwell: Perhaps I can try to summarize where we have got to. We agree that the process of setting RDAs or ULs is approximate because the database isn't there. We agree that setting these numbers has value. In general, my impression from doctors and nutritionists I speak to is that they respect these values. If they actually read the reports and realized how approximate they were they might have a total shock. It is probably not a good idea for them to have total shock, otherwise all kinds of weird nutritional consequences could arise. If we had more more money for research, could we actually do human research, ethically, that is going to improve our database?

Rayman: Our Foods Standards Agency does commission research. This is happening in the USA as well. Our Food Standards Agency is aware of where there are deficits in information and what needs to be done. Some can be done ethically. Such research should be commissioned by governments in my opinion.

Boobis: What can we ethically do in humans? There is a lot we can do without creating frank toxicity beyond the upper level of exposure. Many groups are trying to develop biomarkers of effect. We need to do large studies in humans to validate these biomarkers and understand what the output means. We can do studies on bioavailability, and fate and disposition of supplements. We will be restricted with regards to the duration of exposure issue. We can do focused studies to better understand the relationship between biomarkers of exposure and intake. Again, a huge problem is being able to figure out what the actual exposure is.

Aggett: I think biomarkers are going to be crucial, and we have a whole new raft of opportunities with biomarkers. With intelligent research we might even be able to use and assess biomarkers in vivo in accessible tissues, including the gut. What is really needed is a systematic strategy addressing specific issues for which appropriate biomarkers can be identified. What is sad about nutritional research is that we have a plethora of papers on gross deficiency and a plethora of papers on gross excess, both in experimental models and in humans, but nothing that helps us inform the structure of those thresholds at either end of the plateau of adequacy between inadequate and excess intakes. If one then starts to understand the potential of differential display and post-genomic molecular biology in characterizing metabolic reactions to inadequate and excess intakes there is a lot of scope for a systematic, strategic approach to this. This is a message that we need to get to the funding agencies.

Halliwell: I agree. I'd love to give the IOM some money so it could commission research. The EU is heading in this direction. It is good for industry to come in, but it can't do the whole job, because their funding tends to be patchy.

Russell: What is increasingly being done is partnerships between federal funders and industry support. The ATBC (α-tocopherol/β-carotene) study is one of those models, with NCI and Hoffman LaRoche involved in the study. This is something that we will need to do in the future because the government won't come forth with the huge sums needed for this type of research.

Halliwell: I agree. And in the design of these big studies the biomarker component needs to be built in.

Shekelle: What do people mean when they talk about biomarkers?

Aggett: Interpretation and understanding of biomarkers has expanded recently. There are not only these static makers; one can also use changes in a battery of markers to assess an effect. As long as effects are relevantly related to a hypothesis or clearly defined question, and as long as biomarkers can in turn have their quality assured and their variability expressed in some way, just about anything can be a biomarker.

Rayman: Where biomarkers score is that they make studies a lot cheaper, because studies have a shorter endpoint.

Coates: In the nutrition and cancer field there is not a lot of optimism that there are validated biomarkers that can predict responses. There is enthusiasm which is sometimes is a little ahead of the date of delivery.

Boobis: The biomarker issue gets confused in peoples' minds. Part of this is the hype and enthusiasm. Validation is a key issue. People see a hint and push ahead. We also have to recognize the spectrum of biomarkers. There was one point in clinical studies where people distinguished between a biomarker of effect and a surrogate endpoint. The latter is a mature biomarker that can replace a clinical outcome. Most of the biomarkers we are talking about do not come into this category. If we have a panel of appropriate biomarkers we can look at different facets of the biology and come to meaningful conclusions.

References

Steinberg D 2006 The pathogenesis of atherosclerosis. An interpretive history of the cholesterol controversy, part V: the discovery of the statins and the end of the controversy. J Lipid Res 47:1339–1351

Thacher TD Fischer PT, Pettifor JM et al 1999 A comparison of calcium, vitaminD, or both for nutritional rickets in Nigerian children. N Engl J Med 341:563–568

US Department of Health & Human Services 1996 The 3rd National Health and Nutrition Examination Survey, NHANES III (1988–94). In: NHANES III reference manuals and reports (CD-ROM). Centers for Disease Control and Prevention, Hyattsville, MD

Criteria for substantiating claims

Peter J. Aggett

Lancashire School of Health and Postgraduate Medicine, University of Central Lancashire, Preston PR1 2HE, UK

Abstract. Claims are used to support public health advocacy and marketing. Their evidence base is variable. Claims are made on (i) nutrient content, (ii) comparative merits, (iii) health benefits, and (iv) medical benefits. Experience with therapeutic agents has aided the development of recommendations for the substantiation of health claims for foods and food components, with which dietary supplements would be included. An EU Concerted Activity, Functional Food Science in Europe, suggested that such claims should be based on the general outcomes of 'enhanced function' and 'reduced risk of disease'. A further EU Concerted Activity, The Process for the Assessment of Scientific Support for Claims on Foods, proposed that the evidence base should provide: a characterization of the food or food component to which the claimed effect is attributed; human data, primarily from intervention studies that represent the target populations for the claim; a dose–response relationship: evidence of allowing for confounders including lifestyle, consumption patterns, background diet and food matrix; an appropriate duration for the study; a measure of compliance; and have adequate statistical power to test the hypothesis. When ideal endpoints are not easily accessible for measurement, validated and quality assured markers of the intermediate or final outcomes could be used, as long as their relationship is well characterized. Overall, the totality and coherence of published and unpublished evidence should be considered. Assessments for substantiation need expert judgement, weighting of the strength of the claim, and intelligent use of the criteria applied on an individual basis with respect both to gaps in knowledge and to any need for new knowledge and data.

2007 Dietary supplements and health. Wiley, Chichester (Novartis Foundation Symposium 282) p 46–58

Claims for dietary supplements, foods or food components are widely used in marketing and to support public health advocacy. However, the quality of evidence used to support these claims and their regulation internationally has been variable (Richardson et al 2003). There are probably many reasons for this. Among these are the different kinds of claims that may be made and, until recently, the absence of any internationally agreed systematic code for the evaluation of the evidence on which such claims might be based. Claims are made on (i) nutrient content, (ii) comparative merits, (iii) health benefits, and (iv) medical benefits. The latter are strictly regulated and will not be specifically considered in this contribution.

However, the establishment of a medical claim requires an extensive and robust portfolio of evidence acquired on the basis of hypothesis-led research using well-defined outcomes that are clearly related to the desired medicinal benefit. Thus it is not surprising that experience with the investigation and justification of medical claims should inform approaches to the substantiation of claims for foods and food ingredients.

Recently two European Concerted Actions have specifically explored the evidence base for the functionality of foods and food components (Diplock et al 1999), and the scientific support of any claims made for such products or components (Aggett et al 2005). Both projects involved generic fundamental issues of science that are directly applicable to the substantiation of any functional claims proposed for dietary supplements and this contribution draws extensively on the reports of these Concerted Actions.

A claim is, 'Any representation which states or implies that food has certain characteristics relating to its origin, nutritional properties, nature, production, processing, composition, or any other quality' (Richardson et al 2003). Current initiatives in the categorization of claims within the European Union envisages that claims will relate to 'what the product contains', which will give rise to a nutrient content claim describing it as having a variety of compositional characteristics such as 'low fat', 'fat free', 'no sugar', 'high fibre', 'high in vitamins', 'high in minerals', and so on. The definitions, thresholds and criteria for such claims are based on risk assessments of high and low nutrient exposures and nutrient risk benefit analysis (Renwick et al 2004, Joint FAO/WHO Technical Workshop 2006). The other important category of claim relates to 'what the product does'. This may be one of a range of 'well-established' nutrient functions, many of which are general. Currently regulatory agencies are finalising refinements to the characterization of the types of functional claims that may be made (Richardson et al 2003). Basically, this would enable more specific benefits to be claimed and these would be classified as relating to 'an enhanced function', or to 'a reduced risk of disease'. Any claim relating to the prevention, cure or clinical management of a disease would be a medical claim and is beyond the scope of this commentary.

These developments for foods and food components are clearly relevant to the context of the topic of this meeting, 'Dietary Supplements and Health'. For all such products there is a need to consider the optimal characteristics of the evidence base for any claim that states, suggests or implies that a relationship exists between the food category, a food or one of its constituents and health or well-being: namely 'eating X regularly' does 'something' to the population in general or to a specific subpopulation.

The types of information for substantiating a claim are essentially the same as those for establishing, as much as it can be done, any other scientific truth (Rothman & Greenland 2005). There are epidemiological or natural surveillance data derived from opportunistic or systematic observations that highlight interesting associations between foods or food components and health (or the deterioration of health) and well-being: identify potentially valuable markers or outcomes of a potentially beneficial effect, and in general enable hypothesis generation as the basis of further study. This information may be complemented by a body of data derived from studies *in vitro*, or in animal models. These data would be expected to support and corroborate epidemiological observations, but not always necessarily so because of species differences in metabolism and exposure. Nonetheless, studies in animal models can be designed to refine hypotheses, to elucidate possible mechanisms and to identify markers that could be used in further studies in humans. There are also, in the field of human health, opportunities to use data from secondary intervention studies that provide information of the effect of food or food ingredients on established health conditions.

Collectively these types of information should inform the design and interpretation of specific human intervention studies using definitive outcomes related to the specific claims. These types of studies are not necessarily easy to do. The benefit of the intervention may take a long time (i.e. years) to be effective and the research strategy needs to demonstrate and quantify a sustained exposure to the food or food ingredient, its systemic presence, and outcomes related to the mechanistic pathway leading to the claimed benefit of an enhanced function or reduced risk of disease. This is illustrated in Fig. 1, which is derived from the European Union Concerted Action on Functional Food Science in Europe (FUFOSE) (Diplock et al 1999).

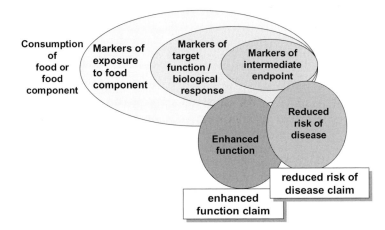

FIG. 1.   The FUFOSE strategic scenario of markers for use in the scientific support of claims for foods (Diplock et al 1999).

FUFOSE was co-ordinated by the International Life Sciences Institute (ILSI) Europe. It was stimulated by the realisation that although the primary role of the diet and its constituent nutrients is to meet an individual's physiological requirements, there are needs and opportunities to change the relative composition of nutrient intakes, the preparation and preservation of foods, and in the organoleptic characteristics of foods in response to changes in lifestyle and consumer expectations. Furthermore, there is an increased appreciation both by scientists and consumers alike that foods and food components can modulate specific functions and thereby have beneficial physiological and psychological effects over and above the customarily accepted nutritional effects. This realization has stimulated research and reconsideration of the regulation of foods and health claims in a number of countries (Richardson et al 2003) and systems have emerged, that are largely similar, to enable the submission of comprehensive evidential databases that would support claims that could be approved by regulatory authorities for use in the labelling of products. FUFOSE assessed critically the science base required to prove that specific nutrients and food components positively affect target functions in the body. It did this from a function driven perspective rather than from a product orientated approach and in this regard it reached a consensus on how specific functional claims might be explored and justified, as well as identifying areas in health and well-being that might benefit from functional foods, and considering appropriate modifications to food and food constituents.

The pathway and markers illustrated in Fig. 1 summarizes an evidence-based approach to nutrition and metabolism (Diplock et al 1999). It demonstrates that the desirable outcome may be removed both in time and pathogenic mechanism from consumption of the food component, and that, in turn, this necessitates the use of intermediate or surrogate markers as 'outcomes'. The identification of these markers requires an informed insight of the relevant pathogenic mechanism and an intelligent selection from what might be a large number of candidate markers. Thus the validation and quality assurance of any selected marker is particularly important if it is going to contribute to the substantiation of the claim (Diplock et al 1999).

Another European Union Concerted Action, also co-ordinated by ILSI Europe, critically assessed the nature of the scientific evidence provided as a support for a claim on foods or food components. This Concerted Action, the Process for the Assessment of Scientific Support for Claims on Food (PASSCLAIM) (Aggett et al 2005), was an extensive precompetitive collaboration involving representative stakeholders such as consumers, health and social care sectors, regulators, legislators, educationalists and industrial experts from throughout Europe. It evaluated existing schemes that assess the scientific substantiation of claims, and through an extensive consultative process developed and tested criteria that could be used to inform the process whereby evidence is assessed by regulatory authorities who are responsible for approving any claim for a food. PASSCLAIM explored existing and opportunities for claims in health and well-being in two phases. The first phase (Asp et al 2003) reviewed bone health and osteoporosis; cardiovascular disease; and physical performance and fitness. The second phase (Asp et al 2004) dealt with body weight regulation, diabetes and insulin sensitivity; diet related cancer; mental state and performance; and gut health and immunity. The Concerted Action proposed criteria that can be used to explore links between diet and health. On that basis it produced a generic guidance tool for assessing the scientific support of health claims on foods and food components (Aggett et al 2005).

The PASSCLAIM criteria (Aggett et al 2005) for the scientific substantiation of claims are as follows:

(1) The food or food component to which the claimed effect is attributed should be characterized.
(2) Substantiation of a claim should be based on human data, primarily from intervention studies, the design of which should include the following considerations:
    (a) Study groups that are representative of the target group
    (b) Appropriate controls
    (c) An adequate duration of exposure and follow-up to demonstrate the intended effect
    (d) Characterisation of the study group's background diet and other relevant aspects of lifestyle
    (e) An amount of the food or food component consistent with its intended pattern of consumption
    (f) The influence of the food matrix and dietary context on the functional effect of the component
    (g) Monitoring of subjects compliance concerning the intake of the food or food component under test.
(3) When the true endpoint of a claimed benefit cannot be measured directly, studies should use markers.
(4) Markers should be: biologically valid in that they have a known relationship to the final outcome, and have a known variability within the target population; methodologically valid with respect to their analytical characteristics.
(5) Within the study the target variable should change in a statistically significant way and the change should be biologically meaningful to the target group consistent with the claim to be supported.
(6) A claim should be scientifically substantiated by taking into account the totality of the available data and by weighing of the evidence.

These criteria reflect an appreciation that any claim for a food, food ingredient or dietary supplement should be based on evidence derived principally from human studies that represent the expected use of the product in the context of customary diet. The criteria also recognize the value of markers of the intermediate effects when ideal endpoints are not accessible measurements. Nonetheless, all the markers used should have a proven validity and be amenable to standard quality assurance. This includes outcomes related to mental health and well-being and to mental and physical performance, for which the activity accepted that a number of objective and valid markers are available. Thus for all the areas reviewed by PASSCLAIM, there are quality assurable markers that would enable the characterization of the effect of any supplement or dietary intervention and the assessment of whether or not the magnitude and character of the effects are statistically and biologically meaningful.

The PASSCLAIM criteria are very much a gold standard. It is unlikely that the financial and temporal resource would be available to substantiate claims for food components or supplements based on the relatively high standards that are embodied within these criteria. However, a proposed claim that is based solely on supporting data from animal studies, or *in vitro* and molecular investigations coupled with observational or anecdotal information would be unlikely to be substantiated in the absence of any intervention study. Whatever the case, the portfolio of evidence would need to be assessed in its totality. As such the collective scientific support for substantiating a claim would need to be addressed with reference to the classic criteria for causal inference, namely the strength of association between a phenomenon and exposure to supplements or food component; the consistency of this association; its specificity and temporal relationship; the dose/exposure-response relationship; the plausibility and coherence of the data; and the availability and quality of other supportive evidence such as that relating to analogous products or components (Hill 1965). This is essentially weighing the totality of the evidence, but the evidence presented to substantiate a claim should not be selectively picked to support a proposed claim.

All relevant published studies, and ideally, but probably not realistically, all unpublished data should be available for the assessment of the scientific support that any claim. Probably the only basis for selective presentation of studies and their outcome should be on the basis of the quality of data, and if this is done, then the bases for exclusion should be explained. Similarly the extrapolation of effects on claims between different population groups, for example between adults and children or men and women should be done transparently and be subject to risk benefit analysis.

Even if the criteria listed here are regarded as idealistic or aspirational, and perhaps difficult to achieve, they should be valued for this. They provide a sound basis for the operation of transparency, objectivity, integrity and quality in the review process. They should thus improve the efficiency of regulatory review. Although these criteria set a standard of the quality of the database and of its use, they do not of themselves establish a claim. The use of the criteria would still need to draw on the broad spectrum of scientific data and it is difficult to establish a definite rule of the justification of the claim. Each claim would need to be assessed in its own right and this would need an intelligent appraisal of the comprehensiveness of the knowledge base and sensitivity to any gaps in it, and to the development of new knowledge that may bridge these. Thus there will still be a need in the advisory and regulatory process for an intelligent interpretation of the evidence and informed independent scientific advice. The PASSCLAIM criteria were drawn up as a yardstick and the process involved deliberately avoided setting up any algorithm or scoring system that might enable a quantitative expression of the strength of the scientific evidence submitted in support of a claim. It was felt that the criteria could serve as a universal standard that would enable international and inter-agency harmonisation of approaches to establishing claims. Given the relatively uncompromising standard set by the criteria it is probably best left to competent authorities who can on a case by case basis perform a risk benefit analysis of any portfolio for a claim. Thus the PASS-CLAIM consensus has made no recommendation on how the evidence should be weighted e.g. graded from A to D, or from 'Convincing' through 'Probable', or 'Possible' to 'Insufficient'. The flexibility of any system to allow or qualify claims should lie with competent authorities acting on the advice of stakeholders. The criteria outlined here offer a basis for scientific integrity. Even though the criteria apply specifically to the assessment of a portfolio of submitted evidence and are not meant to provide a template for a research strategy for any product, it was appreciated and hoped that those who are responsible for compiling and acquiring the scientific support for any claim would be able to use the criteria as a guide.

References

Aggett PJ, Antoine J-M, Asp N-G et al 2005 PASSCLAIM: Process for the Assessment of Scientific Support for Claims on Food: Consensus on Criteria. Eur J Nutr 44 (suppl 1): 1–30

Asp N-G, Cummings JH, Mensink RP, Prentice A, Richardson DP 2003 Process for the Assessment of the Scientific Support for Claims in Foods (PASSCLAIM)-Phase One; Preparing the Way. Eur J Nutr 42 (Suppl 1):1–119

Asp N-G, Cummings JH, Howlett J, Rafter J, Riccardi G, Westenhoefer J 2004 Process for the Assessment of the Scientific Support for Claims in Foods (PASSCLAIM)-Phase Two; Moving Forward. Eur J Nutr 43 (Suppl 2):1–183

Diplock AT, Aggett PJ, Ashwell M, Bornet F, Fern EB, Roberfroid MB 1999 Scientific Concepts of Functional Foods in Europe: Consensus Document. Br J Nutr 81(Suppl 1):S1–S27

Hill AB 1965 The environment and disease: association or causation? Proc R Soc Med 58:295–300

Joint FAO/WHO Technical Workshop 2006 A model for establishing upper levels of intake for nutrients and related substances. World Health Organisation, Geneva

Renwick AG, Flynn A, Fletcher RJ, Muller DJG, Tuijtelaars S, Verhagen H 2004 Risk-Benefit analysis of micronutrients. Food Chem Toxicol 42:1903–1922

Richardson DP, Affertsholt T, Asp N-G et al 2003 PASSCLAIM-Synthesis and review of existing processes. Eur J Nutr 42 (suppl 1):96–111

Rothman KJ, Greenland S 2005 Causation and causal inference in epidemiology. Am J Public Health 95:S144–S150

DISCUSSION

Taylor: You mentioned relative to 'weighing the evidence' that your health claim process stops before you weigh the evidence, on the grounds that you are using scientific judgement. Yet as I listened to the concept of intelligent application, it seems to me that this is scientific judgement too and, as such, a weighing of the evidence. I'm curious about the demarcation between weighing the evidence for the final outcome versus the activities that occur scientifically right up until that final step.

Aggett: It is not that it is inappropriate to weigh the evidence, rather that this should be part of the advisory process, and the evidence should not necessarily be weighed just on the basis of whether or not the criteria are fulfilled. In terms of weighing the evidence in the context of a claim, this would be a broader interpretation.

Yong: In a world where people are taking lots of supplements and botanicals, while no one can doubt the system you proposed, the issue is how we can move the system onward from the current position, with no or little evidence, to one where there is better evidence, and finally to the rigorous system you propose. Can we adopt a system based on the fact that a little evidence is better than no evidence, and then recommendations can be made?

Przyrembel: This has nothing to do with dietary recommendations. We are dealing here with claims that people want to make to sell their products. If people are not allowed to make health claims, they can still sell their product; they just can't make a claim. The idea that the criteria for making claims should be very strict is one I approve of.

Manach: I wanted to insist on the importance of a good description of the product you want to make a claim for. For example, with polyphenols you may have quercitin in your product, but this occurs in forms with rather different bioavailabilities. One form can be absorbed and the other might not, leading to conflicting results.

Aggett: One of the important things is to make sure that people don't extrapolate from one polyphenol to another without justification.

Manach: For example, there are many products produced from soy alfaflavins. Some of them are produced from the germ, others from the whole grain. The composition of the products is very different.

Aggett: And sometimes it varies according to the manufacturer's individual process. In the area of probiotics for example, there is a whole mixed bag of evidence. The absolute and relative bacterial colonies in the same product may be inconsistent. They are not always shown to be viable. In experimental studies they may be tested on one occasion as an aqueous suspension and another time they are tested in a milk ferment. However, the companies, or a company, may be trying to justify a single claim for the probiotic, extrapolating from these data without any reference to the possible confounding effect of the product's or dietary matrix.

Alexander: You have addressed the possible adverse effects and safety of these products. They are beyond those for vitamins and minerals where there are upper limits (ULs). This applies to bioactive compounds in food in general. There is the potential for adverse effects as well. We have seen a lot of promotion of phyto-oestrogens, for example: the producer seems to believe that they are completely without side effects, which I doubt.

Aggett: In this context of setting out the criteria for claims, part of the preamble assumes that the components would have been reviewed under existing processes for safety evaluation. Earlier we mentioned risk–benefit analysis and reference was made to niacin and flushing. If people want to market something containing niacin on the basis of reducing cholesterol, the levels needed would be sufficient to cause flushing in a third of the population.

Katan: The report you produced was valuable and I hope the EU will act on it. I wanted to address the variety of claims printed on a package in order to sell it. The reason a consumer buys a product is not because it will increase their stool weight but because it will make them feel better or prevent a disease. Many companies are trying to get into the lower type of claims (which are easier to get away with) while at the same time suggesting an effect of a higher type. Wouldn't it be better if we had just one sort of claim (namely that a supplement increases health and well-being), and judge claims on this basis, not allowing companies to get away with stating something like 'this product is rich in Ca^{2+}, and Ca^{2+} is needed for strong teeth.' We know that eating more Ca^{2+} has no effect on your teeth, but the consumer does not know this.

Aggett: I have a lot of sympathy with this. The EU is still driven very much by commercial imperatives. Some of us have already been accused of 'pharmaceuticalizing' the food industry, forcing it into a pharmacological-type model.

Przyrembel: The EU will collect these types of claims and send them to the European Food Safety Authority (EFSA), which will have to authorize them. One of the most widely made claims is that Ca^{2+} is good for teeth and bones.

Azzi: Your presentation was outlining a kind of 'constitution'. How can you transform the constitution into law? I was thinking of the Thomson-ISI impact factor, ranking the scientific journals as a function of the citations that articles have received. It would be nice to find an easy way to attribute an 'impact factor' to different claims. It would be good to come up with a simple value judgement and to make it clear to the public.

Aggett: We might be toying with doing something like that. When it comes to cardiovascular systems, some of the working groups were impressed that cholesterol and blood pressure are predictably associated with outcomes. There is consideration of looking to see if we can evaluate or identify the biomarkers that are being used at the moment. The feeling is that the old, established markers are of as much benefit as the more sophisticated molecular markers, simply because the old markers have been validated to an extent.

Shekelle: When we were doing our analysis of the ephedra data we wrestled with the following question: should the level of evidence to substantiate a claim of benefit be different from that required to be worried about harm, particularly if the harm is devastating? If we are talking about pharmaceuticals, to say something has a benefit means you have to cross the threshold of an alpha error rate. We have this convention of $P = 0.05$. If you are worried about harm, do you also have to prove that this harm is unequivocally related to this input with a one-in-20 chance, or is some lesser confidence in the harm acceptable? This is not a small issue. The benefit of these supplements is usually relatively modest, and their safety is pretty good, but there is always the potential for rare harms. What level of evidence do we need to be able to deal with these?

Aggett: One of the references Alan Boobis gave was a risk–benefit analysis by Renwick et al (2004). It didn't come up with an answer, but it showed how one could construct an approach to try to assess a certain level of intake above which there was no further benefit, and to assess at that level, what percentage of the population might have a manifestation of an adverse effect.

Boobis: I always get uneasy when I hear questions like that, because of the precautionary principle. Either something is causing harm or it is not. What we have in this case is equivocal evidence that it is causing harm. If we have no clear evidence that it is causing harm, then why are we still concluding that it is likely to cause harm? If applied generally, we would end up accepting nothing as safe.

Shekelle: I understand what you are talking about in terms of a decision analysis approach. The ephedra case that we wrestled with is the following: randomized evidence of weight loss as an intermediate outcome, with the usual levels of conventional statistical significance, and in the same studies a variety of minor nuisance side effects. Then there were a lot of case reports of death. There was a case control study for stroke which had a P of 0.07. 3000 people have been randomized in all the trials done to date. The manufacturers say there is no evidence that ephedra causes stroke or death because it is not proven. How do you deal with this situation?

Katan: The present paradigm in Europe is that even the faintest whiff of risk would be enough to ban a food component. My opinion is that if the substance really helps where there is no other source of help, such as a dietary supplement with a proven benefit for irritable bowel syndrome, you should accept some side-effects. On the other hand, if you are talking about plant sterols for lowering cholesterol, any evidence of harm should be enough to ban them because we already have excellent and safe drugs for lowering cholesterol.

Przyrembel: There is some imbalance in discussions of the evidence for benefits and the proof of safety. One should not forget that dietary supplements are foods, so the level of evidence for their safety should be much higher than for drugs. Drugs are given to people with a disease and some risk of adverse effect is acceptable in view of the benefit expected. This is not the case with dietary supplements.

Boobis: I don't disagree: the view has been taken that the hurdle for proving benefit is lower than the hurdle for proving absence of harm. What I am concerned about is that we don't use evidence for an absence of harm and then twist this into 'we just don't know'. This depends on the power of the study. We get a statistical level of 0.07 that shows there is no significant evidence of an increase in stroke incidence, but that is enough to say that we are concerned. If you add to this case studies with reasonable evidence of exposure and possible association with the outcome then you might build up a negative picture. Where do you draw the line?

Shekelle: This exact issue came up and we concluded that there probably was a causal relationship, although it was not proven at the conventional levels of statistical significance.

Boobis: If you go to the epidemiological literature you will find relative risks of less than 2 (e.g. 1.5, 1.3) for a number of environmental exposures, but which are still significantly positive at $P < 0.05$. It can still be argued that this is evidence that there is a problem.

Wharton: How do you think the EU will handle this report? I have heard some suggestion that people will have to submit their claims to a review body to see if it agrees. This is quite unlike any other food legislation in the EU. Normally the EU lays down a directive that the member countries have to comply with. A review process concerning composition of foods doesn't exist in any other directive, to my knowledge. Coming to contemporary examples, where are we with the claims about large B6 intakes and premenstrual tension? There are many claims about ω-3 fats affecting childrens' learning ability: would these still be allowed? A leading food manufacturer makes claims about probiotic yoghurts modulating the immune system. Are these still OK?

Aggett: The issue is that currently they are not disallowed. The intention is to try to induce some rationality into the field, so that one knows whether or not these are genuine claims. At the moment there are different systems in most of the EU member states. Modulating the immune system is a claim used in the UK, but in Sweden this is under challenge at the moment. The claims made about ω-3 fatty acids and intellectual performance was an intriguing issue. At one point this claim came under review as part of the case for including ω-3s in infant formulas. Its failure to convince the review panel meant that its inclusion was deferred, even though it was included in the formulas for pre-term infants. The situation in the UK with B6 is that the expert group on vitamins and minerals, convened as a response to some of the controversy over a review of a safe upper level of intake for B6, actually supported the recommendation on the basis of neuropathic tingling phenomena occurring in the women who were consuming it. It perhaps is an example where some modulation of the uncertainty factor from animal-based data enabled the expert group to come up with a value similar to that derived by the COT. This is five times lower than the value set in the USA for B6. You also asked where we were in the context of European legislation, perhaps Hildegard can comment on this.

Przyrembel: The claims directive is not yet finalized. There will be two parts in it: a register with generally accepted claims (member states send in their claims, EFSA will assess them, and, on approval they will be in a register of claims that everyone can use). In addition, there will be a register of authorized disease risk reduction claims. They are to be individual claims made for a specified product and evaluated by EFSA. But this latter regulation is under lively discussion at the moment in the European parliament because some groups in parliament want only a notification procedure: the manufacturer wants to make a claim and notifies the authorities, and only if someone knows that this claim is really wrong, then the authorities could take action. The European Commission wants an authorization procedure with an evaluation by EFSA.

Halliwell: It is an interesting polarization.

Yetley: I'd like to return to the issue of whether the same kind of evaluation is applicable for harmful effects as it is for beneficial effects. One of the challenges dealing with supplements and other nutritional products is that we frequently don't have a systematic battery of toxicology testing prior to doing studies of benefit or when applying effects of benefit to practical situations. Generally, the safety data come from studies of benefit that have been designed for purposes other than evaluating safety. For example, they are generally not powered to evaluate safety; nor have the inclusion/exclusion criteria or randomization protocol been developed to evaluate safety. So the question is often asked as how best to interpret and weigh safety data from studies not designed to evaluate a safety endpoint.

Ernst: That is not entirely correct. Kava, a herbal anxiolytic marketed as a food supplement, was banned on the basis that it had efficacy for anxiolytic properties from about a dozen controlled trials which showed that its adverse effect rate was like placebo, but what killed it were case reports, first from Switzerland, then from Germany, then from around the world. By the time it was banned there were just 70 case reports.

Reference

Renwick AG, Flynn A, Fletcher RJ, Muller DJG, Tuijtelaars S, Verhagen H 2004 Risk–Benefit analysis of micronutrients. Food Chem Toxicol 42:1903–1922

Science in the regulatory setting: a challenging but incompatible mix?

Elizabeth A. Yetley

Office of Dietary Supplements, National Institutes of Health, 6100 Executive Blvd., Rm. 3B01, MSC 7517, Bethesda, MD 20892-7517, USA

Abstract. Regulatory decisions informed by sound science have an important role in many regulatory applications involving drugs and foods, including applications related to dietary supplements. However, science is only one of many factors that must be taken into account in the regulatory decision-making process. In many cases, the scientific input to a regulatory decision must compete with other factors (e.g. economics, legal requirements, stakeholder interests) for impact on the resultant policy decision. Therefore, timely and effective articulation of the available science to support a regulatory decision can significantly affect the relative weight given to science. However, the incorporation of science into the regulatory process for dietary supplements is often fraught with challenges. The available scientific evidence has rarely been designed for the purpose of addressing regulatory questions and is often preliminary and of widely varying scientific quality. To add to the confusion, the same scientific evidence may result in what appears to be different regulatory decisions because the context in which the science is used differs. The underlying assumption is that scientists who have a basic understanding of the interface between science and policy decisions can more effectively provide scientific input into these decisions.

2007 Dietary supplements and health. Wiley, Chichester (Novartis Foundation Symposium 282) p 59–76

Regulatory decisions benefit by being fully informed by sound science. The incorporation of science into regulatory decisions for dietary supplements is a logical sequence to the science/policy interactions used for conventional foods and drugs. However, differences in product characteristics, consumer use patterns, regulatory frameworks and the nature of the available scientific evidence present challenges for effectively incorporating science into the supplement regulatory processes.

This article focuses on the interactions between science and legal/regulatory policies for dietary supplements. An underlying assumption is that scientists who have a basic understanding of the interface between science and policy decisions can more effectively provide scientific input to legal and regulatory processes. By necessity, the examples and descriptions of the regulatory decision-making processes derive from the author's knowledge of the US regulatory system. However, to the extent possible, the interactions of science and policy decisions are presented in a manner that has generalizability across legal and regulatory systems in other countries.

Context matters

Who bears the burden?

When is the manufacturer responsible for collating and evaluating the evidence on safety and effectiveness and when does the regulatory agency bear this responsibility? The answer to this question affects the ability of the regulatory agency to provide public assurance that a product that enters the marketplace is safe and truthfully labelled. It also affects the ability of the regulatory agency to achieve timely action against unsafe or mislabelled products once they are in the marketplace.

Generally, evaluations of safety and effectiveness occur at two points in time. The first is prior to marketing of the product (Fig. 1). Here the manufacturer bears the burden for summarizing the evidence. The second time involves evaluation of safety and effectiveness after the product is marketed. Once a product is marketed, the regulatory agency, not the manufacturer, bears the burden for collating and evaluating the evidence if concerns arise about the safety or labelling of products in the marketplace.

This pre- and post-market split in responsibility for bearing the burden of scientific evaluations is generally the same for all drug and food products. Differences in regulation of these products are affected instead by whether or not the manufacturer is required to submit his/her pre-market scientific evidence to the regulatory agency for review and approval prior to marketing. Differences also derive from differences in the legal definitions and substantiation standards for safety and effectiveness.

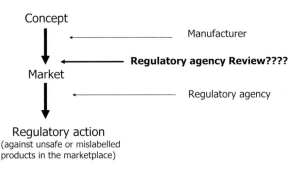

FIG. 1.   Who bears the burden?

Why does it matter whether or not the regulatory agency has access to the manufacturer's pre-market evidence? This access affects the agency's ability to conduct post-market evaluations of safety or effectiveness in a timely manner. For example, for food additives added to conventional foods, the manufacturer must submit to the regulatory agency the evidence for his/her conclusion that the ingredient is safe prior to marketing. On the other hand, for most supplements, the manufacturer is not required to submit to the regulatory agency his/her evidentiary basis for concluding that a supplement ingredient is safe prior to marketing. If, once in the marketplace, the regulatory agency wishes to evaluate the safety of a supplement ingredient, it must *de novo* collate and evaluate the publicly available evidence. The product remains in the marketplace while the regulatory agency conducts its evaluation, which can take considerable time.

Ephedra is an example of the US Food and Drug Administration (US FDA) bearing the burden to establish that a marketed dietary supplement ingredient is unsafe. The agency began the process of investigating the safety of ephedra after numerous reports of possible adverse effects associated with use of these products. The agency bore the burden of collating and evaluating the evidence while ephedra-containing products remained in the marketplace. The agency first convened an advisory committee to evaluate reports of possible adverse effects in 1995 (FDA 1997) and published final regulations banning the use of ephedra in supplement products in 2004 (FDA 2004). The agency's decisions were aided by the availability of a comprehensive evidence-based systematic review (Shekelle et al 2003) that was initiated after possible safety concerns about marketed products had been raised.

What is the question?

The legal or regulatory context—that is, the starting question and the legal standard and process that must be addressed by the available evidence—can significantly affect ultimate regulatory decisions. With different legal frameworks and processes as well as different substantiation standards among types of products, the same or similar scientific evaluations on a given topic can result in quite different regulatory decisions. Figure 2 shows how the nature of the starting question can vary. For example, if the starting question or assumption is that a product or ingredient is unsafe, then this assumption can only be rejected if there is sufficient evidence to demonstrate that the legal standard for establishing safety has been met. Similarly, for claims of effectiveness where the starting assumption is that the product or ingredient is ineffective, this assumption prevails until there is sufficient evidence to support a conclusion that the legal standard for effectiveness has been met. Conversely, if the beginning assumptions are that the product is safe and effective, evidence is needed to reverse these assumptions.

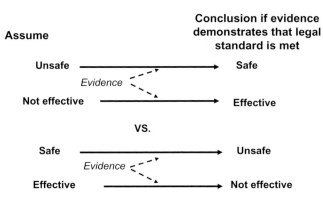

FIG. 2.   What is the starting question?

An example of how the same scientific evidence can result in different regulatory outcomes can be illustrated by safety evaluations of stevia (*Stevia rebaudiana* Bertoni) for use in conventional foods or in dietary supplements. The US FDA (FDA 1995, 1996a) did not object to the use of stevia in dietary supplements because the very limited evidence available at the time was inadequate to demonstrate that stevia would pose 'a significant or unreasonable risk'—the legal standard for evaluating the safety of dietary supplements. Conversely, the US FDA objected to the use of stevia as a food additive in conventional foods because the same very limited evidence was insufficient to demonstrate 'a reasonable certainty of no harm'. (The Scientific Committee on Food of the European Commission also found the evidence to be insufficient to conclude that stevia could safely be used in novel foods [SCF 1999]). In the US FDA example, differences in starting assumptions and legal standards resulted in different regulatory decisions for use of stevia in conventional foods vs. dietary supplements—despite the fact that the same scientific evidence was used for both situations. Generally, the same body of evidence is most likely to result in differences in regulatory decisions when the evidence is weak and inconclusive—as was the case with stevia. Alternatively, a relatively strong body of evidence is more likely to result in similar regulatory decisions under different decision-making contexts.

What is the nature of the available evidence?

Product-specific vs. generic evidence.   Because dietary supplements are regulated as foods, the evaluation of safety and effectiveness is frequently done on a generic ingredient basis rather than on a product-specific basis as is done for drugs. The published scientific literature constitutes most, if not all, of the available data. These studies are generally designed for different purposes than that of addressing regulatory questions. They lack a systematic approach to answer a specific regulatory question because they represent investigator-initiated studies rather than a targeted series of studies. Thus, the available evidence for a particular regulatory decision about dietary supplements is often characterized by significant data gaps, studies of widely varying scientific quality, and questionable relevance. Scientific judgment then becomes necessary to deal with uncertainties associated with extrapolations from studied groups to the groups that are the primary target group for a supplement product or claim and from studied doses to appropriate intakes for dietary supplements.

For example, the 1996 authorization of a health claim on the use of diets rich in folic acid and reduced risk of neural tube defects (NTDs) was based on a process of combining relevant information from several studies (FDA 1996b). None of these studies directly addressed the question at hand as to whether folic acid alone at intakes acceptable for foods and dietary supplements (i.e. ≤400 μg/d) was effective in reducing the risk of pregnancies affected by NTDs in the general population of women of child-bearing age. Strong evidence of a causal relationship between folic acid alone and reduced risk of NTDs was obtained from a trial that evaluated the effectiveness of several single nutrients (MRC 1991). However, the relevance of these findings to the target population of women of child-bearing age was unknown because the study participants were high risk women with a prior history of NTD-affected pregnancies. Additionally, the dose used was 10 times the US Recommended Dietary Allowance (RDA) and so the effectiveness of folic acid at dietary levels was unknown. To assess the likelihood of effectiveness for the general population, results from a trial using a folic acid-containing multivitamin/mineral in the general population of women without a prior history of NTDs were used (Czeizel & Dudas 1992). However, the population was European and relevance to US women and US baseline folate intakes was not known. The likelihood that intakes appropriate for supplement products would be effective in US women was supported by results from observational studies showing an association between daily use of multivitamin/mineral supplements containing 400 μg folic acid and reduced risk of NTDs (Werler et al 1993). Although none of these studies was done specifically for the purpose of evaluating a food and supplement health claim and none directly addressed the regulatory question, an expert advisory committee used scientific judgment to reach the conclusion that there was a sufficient basis from these and other relevant data to support a health claim. Subsequent studies have confirmed the validity of these early scientific reviews (Picciano et al 2007).

Absence of dose–response data. As noted above, evaluations of dietary supplement safety and effectiveness are often dependent on studies designed for other purposes. One notable deficiency is the frequent lack of dose–response data, particularly at intakes of interest for supplement products. To control study costs, intervention trials frequently use only one or two doses—usually at high levels to minimize the likelihood of false negatives. Dose ranging studies prior to designing Phase III trials have not often been done for nutrient- and food-related interventions although there is a trend to change this practice. Thus, while dose–response curves are desirable for evaluating safe and effective intakes for supplement products, these data are often lacking. Consequently, expert scientific judgment is required to decide if and how the available data can be combined and extrapolated to intake conditions consistent with use of supplement products.

Safety vs. benefit: lack of balance between databases. By its very nature, ethical considerations generally preclude a direct causal demonstration of safety through the use of controlled human trials. For conventional food additives, a systematic battery of toxicological testing in several animal models is conducted to identify potential adverse effects. For drugs, pre-clinical screening for possible adverse interactions with other drugs is also untaken. For most nutrients and other ingredients used in dietary supplements, systematic toxicological testing is rarely done (EVM 2003, WHO/FAO 2006). Therefore, safety information must be inferred from studies designed to evaluate benefit or to identify mechanisms of action for essential nutrients. Because these studies are not designed for the purpose of evaluating safety, their ability to detect potential safety concerns is often limited.

For example, inclusion/exclusion criteria for participants in intervention trials are likely designed to eliminate persons at risk. Power calculations are based on the need to demonstrate effectiveness and are often inadequate for detecting statistically significant differences on safety endpoints. Study durations may be too short to detect safety concerns. Conversely, for evaluations of potential beneficial effects of dietary supplement ingredients, direct observations of benefit can be determined with the use of intervention trials. If the current practice of conducting trials of benefit with a general absence of systematic pre-clinical toxicology and Phase I and II clinical testing continues, it will perpetuate the imbalance in the strength, weight, and compelling nature of the evidence for evaluating questions of benefit as compared to questions of safety.

Interaction between science and policy

Multiple opportunities for input

There is generally ample opportunity for scientific input at various stages of the legal and regulatory processes. Unfortunately, scientists do not always take advantage of these opportunities thereby leaving a void to be filled by other stakeholders who may not place a high priority on scientific and public health perspectives (Fig. 3).

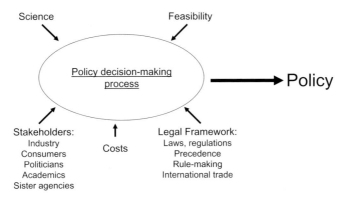

FIG. 3.   Policy decision-making process. Here, 'policy' refers to legislation or regulations. Science is only one of many factors that impact on a given policy decisions. However, you can enhance the likelihood that sound science will be a major influencing factor by ensuring that: the scientific conclusions and recommendations are relevant and appropriate for the context in question, and the science is rigorous and meets generally accepted standards of scientific quality and evaluation.

As with all food and drug regulation, the process of regulating dietary supplements involves two steps. First is the passage of legislation to create the legal framework and to identify the responsible regulatory agency. Second, the responsible regulatory agency develops regulations and implements associated enforcement policies. Both steps provide multiple opportunities for scientific input.

The first step is the passage of laws by legislative bodies. These laws often include specification of scientific standards for substantiation of safety and claims and also delegate (or fail to delegate) associated implementation and enforcement authorities to the designated regulatory agency. For example, the 1994 Dietary Supplement Health and Education Act (DSHEA 1994) changed the substantiation standards and process for demonstrating safety of supplement ingredients. DSHEA also authorized the use of claims that were outside the required pre-authorization process. These changes affected how science interacts with the regulatory process and the ability of the responsible regulatory agency (US FDA) to access and review the available evidence for safety and effectiveness decisions.

The second step is implementation of the legislation by the regulatory agency through the development of regulations and guidance. These processes usually involve solicitation of public comment by the regulatory agency. They may also involve the convening of external advisory committees although this is the exception rather than the rule.

In both the development of legislation and the implementation of that legislation by the regulatory agency, science is only one of many inputs (Fig. 3). To the extent that strong and well articulated science is brought to the table, the likelihood of science influencing the ultimate decisions is enhanced. However, if the scientific community remains relatively uninvolved or ineffectively articulates their perspectives, then science is likely to have less impact on the ultimate decisions.

Invited vs. self-initiated scientific inputs

Scientists have several different ways in which they can interface with the legal and regulatory decision-making processes described above. These include participation in advisory committees and the solicitation of individual expert opinions by the legislative bodies or regulatory agencies. However, scientists can also provide input through self-initiated comments at times when public comment is allowed or solicited.

Invited participation. A type of advisory committee generally familiar to scientists is one that is designed to work independently of persons and groups that may be impacted by a particular regulatory decision (e.g. the regulatory agency itself, industry stakeholders, consumer advocacy groups). Independent advisory committees are a central component of risk assessments including nutrient risk assessments (WHO/FAO 2006, EVM 2003), the setting of nutrient intake reference values (IOM 2001), and scientific consensus conferences (NIH 2005). The independence of advisory committees lends credibility to the science used in regulatory decisions (IOM 1992). To maintain scientific integrity and credibility, committee membership is usually limited to recognized experts in a technical or scientific field that is relevant to the issues under consideration. The credibility of conclusions from an independent advisory committee can be seriously compromised if the committee strays from its scientific focus and adds policy comments to their scientific conclusions. A mixing of science and policy may create an impression that the committee's scientific deliberations were manipulated to support the outcome wanted.

A second type of advisory committee can be convened to provide advice to the regulatory agency at the beginning of a regulatory process by helping to define the science-based questions that need to be answered within the larger regulatory context. Additionally, this type of advisory committee may be convened after the regulatory agency has drafted tentative policy positions—to provide advice on whether the use of the science by the regulatory agency was appropriate within the larger regulatory context at hand. Unlike the independent scientific assessments noted above, this type of advisory committee can appropriately mix science and policy discussions because it is evaluating the appropriateness and practicality of the application of science within a given regulatory context (Fig. 3). In addition to scientific expertise, this type of committee often includes stakeholders with expertise for evaluating cost and feasibility options and who have knowledge of consumer or industry concerns.

General solicitation for public comment. An impactful, but rarely used, opportunity for input by scientists into legal and regulatory decisions is for scientists and scientific organizations to be actively involved in talking with legislators as laws are being developed and in providing responses to regulatory agency solicitations of public comment for proposed regulations and guidance. For example, although many scientists now criticize what they perceive as inadequate public health and science-based provisions of the Dietary Supplement Health and Education Act of 1994, individual scientists and scientific organizations generally remained silent during Congressional considerations and development of this legislative framework. Additionally, despite the solicitation of public comment by regulatory agencies on draft regulations and guidance with relevance to dietary supplements, substantive comments from qualified scientists are infrequently received in response. Silence or ineffective comments at these significant decision points mean that sound science and associated public health considerations may be less impactful than will other stakeholder interests (e.g. marketing and trade considerations, consumer advocacy positions).

Summary

Sound science is a highly desired input into legal and regulatory decisions for dietary supplements and other types of products which can impact public health. Science inputs, however, must effectively compete with inputs from other stakeholder interests, e.g. costs, feasibility, consumer interests, manufacturer burdens. The impact of science can be enhanced if scientists understand the importance of their inputs and also understand the contexts in which their inputs interface with decision-making processes. This understanding can increase the likelihood that science will play an important role in decisions related to dietary supplement regulation.

References

Czeizel AE, Dudas I 1992 Prevention of the first occurrence of neural-tube defects by periconceptional vitamin supplementation. New Engl J Med 327:1832–5

DSHEA (Dietary Supplement Health and Education Act) 1994 Internet access: http://www.fda.gov/opacom/laws/dshea.html (accessed 25 March 2007)

EVM (Expert Group on Vitamins and Minerals) 2003 Methods. Chapter 3 in Safe Upper Levels for Vitamins and Minerals: Report of the expert group on vitamins and minerals. Food Standards Agency Publications ISBN 1-904026-11-7, United Kingdom. Internet access: http://www.food.gov.uk/multimedia/pdfs/vitmin2003.pdf (accessed 25 March 2007).

FDA (Food and Drug Administration) 1995 Letter of Aug 16, 1995, from Linda S. Kahl, FDA, to W. Patrick Noonan, Sunrider Corp. (See Code M2 for *Stevia rebaudiana*, Bertoni (Stevia or stevia leaf). In: Table of New Dietary Ingredient Notifications). Internet access: http://www.cfsan.fda.gov/~dms/ds-ingrd.html (accessed 25 March 2007)

FDA (Food and Drug Administration) 1996a Automatic detention of Stevia leaves, extract of stevia leaves, and food containing stevia. Internet access: http://www.fda.gov/ora/fiars/ora_import_ia4506.html (accessed 25 March 2007)

FDA (Food and Drug Administration) 1996b Food Labeling: Health Claims and Label Statements; Folate and Neural Tube Defects. Final Rule. Fed Register 61(44):8752–8781, March 5, 1996. Internet access: http://www.cfsan.fda.gov/~lrd/fr96305a.html (accessed 25 March 2007)

FDA (Food and Drug Administration) 1997 Dietary Supplements Containing Ephedrine Alkaloids; Proposed Rule. Fed Register 62(107):30678–30724, June 4, 1997. Internet access: http://www.fda.gov/OHRMS/DOCKETS/98fr/97-14393.pdf (accessed 25 March 2007)

FDA (Food and Drug Administration) 2004 Final Rule Declaring Dietary Supplements Containing Ephedrine Alkaloids Adulterated Because They Present an Unreasonable Risk; Final Rule. Fed Register 69(28):6788–6854, Feb 11, 2004. Internet access: http://www.fda.gov/OHRMS/DOCKETS/98fr/04-2912.pdf (accessed 25 March 2007)

IOM (Institute of Medicine) 1992 Food and Drug Administration Advisory Committees. Committee to study the use of advisory committees by the Food and Drug Administration. National Academy Press, Washington DC

IOM (Institute of Medicine) 2001 Dietary Reference Intakes for vitamin A, vitamin K, arsenic, boron, chromium, copper, iodine, iron, manganese, molybdenum, nickel, silicon, vanadium, and zinc. A Report of the Panel on Micronutrients, Subcommittees on Upper Reference Levels of Nutrients and of Interpretation and Uses of Dietary Reference Intakes, and the Standing Committee on the Scientific Evaluation of Dietary Reference Intakes. National Academy Press, Washington, DC. Internet access: http://www.nap.edu/catalog.php?record_id=10026#toc (accessed 25 March 2007)

MRC Vitamin Study Research Group 1991 Prevention of neural tube defects: results of the Medical Research Council Vitamin Study. Lancet 338:131–7

NIH Consensus Development Program 2005 Internet access: http://consensus.nih.gov/ABOUTCDPPDF.pdf (accessed 25 March 2007)

Picciano MF, Yetley EA, Coates PM 2007 Folate and health. Perspect Agric Vet Sci Nutr Nat Resour 2:1–18

SCF (Scientific Committee on Food) 1999 Opinion on *Stevia Rebaudana* Bertoni plants and leaves. European Commission, CS/NF/STEV/3 Final. 17 June 1999. Internet access: http://www.food.gov.uk/multimedia/pdfs/stevioside.pdf (accessed 25 March 2007)

Shekelle P, Morton S, Maglione M, et al 2003 Ephedra and Ephedrine for Weight Loss and Athletic Performance Enhancement: Clinical Efficacy and Side Effects. Vol. 1: Evidence Report and Evidence Tables and Vol. 2: Appendixes. Evidence Report/Technology Assessment No. 76 (Prepared by Southern California-RAND Evidence-based Practice Center, under Contract No 290-97-0001). AHRQ Publication No. 03-E022. Rockville, MD: Agency for Healthcare Research and Quality. March 2003. Internet access: http://www.ahrq.gov/downloads/pub/evidence/pdf/ephedra/ephedra.pdf (accessed 25 March 2007)

Werler MM, Shapiro S, Mitchell AA 1993 Periconceptional folic acid exposure and risk of occurrent neural tube defects. J Am Med Assoc 269:1257–1261

WHO/FAO (World Health Organization/Food and Agriculture Organization) 2006 A Model for Establishing Upper Levels of Intake for Nutrients and Related Substances. Report of a Joint FAO/WHO Technical Workshop on Nutrient Risk Assessment, Geneva, Switzerland 2–6 May 2005. WHO Press, World Health Organization, Geneva, Switzerland. Internet access: http://www.who.int/ipcs/methods/nra/en/ (accessed 25 March 2007)

DISCUSSION

Russell: With regard to supplements, there is no requirement by regulatory agencies to demonstrate bioavailability of the supplement! Shouldn't that be part of the process? A scientist was planning to use a very popular supplement in a trial, but the absorption of some of the minerals, particularly zinc, from the supplement was very, very low.

Yetley: Bioavailability is a challenging issue for policy applications because bioavailability is affected by a lot of things other than product characteristics (e.g. the nutritional status of the consumer). While taking bioavailability into account may be important for research settings, the challenge is how best to apply it in policy settings.

Halliwell: How is the EU looking at these issues?

Przyrembel: The dietary supplement directive isn't complete yet. This is one of the problems. It contains a definition and rules about the labelling, and gives a selection of nutrients that may be used. There is also given a list of substances that may be used as sources of those nutrients. Provisions are made that other substances may be used for the time being under the rules of specific member countries until the European Commission takes a final decision. The Commission will have to report by July 2007 to the European parliament about the advisability of regulating the use of these other substances in dietary supplements, but I don't know who is to provide the data on these. There is a responsibility of the manufacturers to make sure that their product is safe. But there is no clear cut process for assessing safety. At the moment safety is assessed according to rules specific to each country. However, if a supplement is legally on the market in one country, it can legally be exported to all the other EU member countries, on the basis that what is legal in one country of the European Community should be legal also in another. It is a complicated system.

Yong: Has there been any improvement on the DSHEA (Dietary Supplement Health and Education Act) for food supplements?

Taylor: DSHEA remains unchanged, if that is your question.

Yong: How do the regulatory bodies intend to encourage the industry to make better products?

Taylor: In the preamble to the 1994 DSHEA legislation, it is made clear that a key purpose of the legislation is to make such products more readily available. Prior to the DSHEA legislation, the regulation of dietary supplements was beginning to be addressed using, in some cases, the same framework as that used for food additives. DSHEA most certainly left dietary supplements under the umbrella of food regulation but it precluded the substances from being reviewed as food additives which can be a lengthy and thorough process. Products could more readily be brought to market, and certainly in the last 10–15 years they have.

Yetley: The same data may be used to frame supplement-related arguments in different ways. For example, in the Institute of Medicine's (IOM) Diet and Health Report (IOM 1989), the IOM suggested that their diet and health recommendations were unlikely to be relevant to dietary supplements because the available data were based on studies relating foods and dietary patterns to reduced risk of chronic diseases. Studies relating single nutrients and dietary supplements were generally unavailable. Generalizing results from food and diet studies to individual nutrients or supplements was deemed problematic. Prior to DSHEA, the FDA published a proposed rule for health claims that cited the 1989 IOM report concerns about the difficulty in generalizing from studies on foods and diets to dietary supplements. In 1994, the preamble to DSHEA cited diet and health relationships consistent with those in the IOM report as the basis for justifying why ready access to dietary supplements was important for public health.

Rayman: I have a comment about regulatory issues in the EU, about which I had some experience in relationship to selenium yeast. Selenium yeast has been on sale in various countries for a long time with no apparent adverse effects. But it wasn't one of the forms of selenium that was defined as being appropriate. There is a chosen group of forms of selenium permitted. Inorganic selenium was defined as being OK, but organic selenium wasn't. Any country that wanted to continue to use selenium yeast had to ask for derogation before a certain date. If they didn't do this, they aren't allowed to continue to sell selenium yeast.

Przyrembel: In Germany we refuse selenium yeast in food supplements as long as it is not clearly defined. The problem with selenium yeast in Germany was that there were very different types on the market, with very different bioavailability.

Rayman: Yes, the selenium yeasts on the market can be variable. But those from large manufacturers are pretty standard. What then happened is that each of the companies making it who wanted to be able to sell their product after 2010 had to put in a dossier, which in one case I know of cost a small company £250 000. Other companies have been hit with the same sorts of bills. It has been a massive job for people who were selling a good quality, inexpensive form of selenium.

Katan: When DSHEA was passed by the US Senate we were all worried about the consequences. It looked like anyone could now sell anything with any claim they wanted. Has anyone tried to quantify the possible damage? Have there been major problems?

Yetley: Ephedra is an example where FDA identified possible safety concerns. The evidence-based review that Paul Shekelle's group did provided a systematic and comprehensive summary of the available data on safety and efficacy.

Katan: Apart from official legal action, does anyone have a feeling if, and what, harm has occurred?

Ernst: You need to define 'harm' broadly. If you consider the estimate that $50 billion dollars have been spent on ineffective treatments and supplements, to most people this would constitute some harm!

Katan: It is hard to say: it could give some people satisfaction to know that they are doctoring themselves in some way.

Przyrembel: There are ideas about collecting reports on adverse effects of dietary supplements. In our institute we are getting spontaneous reports only and mostly via the system for reporting adverse effects of drugs. We identify the products as dietary supplements and file them, and I think in the coming years we will have more information on this. My impression is that the reports on adverse effects of dietary supplements are on the increase. But we must remember that these are reports on adverse effects, and that they do not necessarily mean harm caused by the supplement.

Coates: There is a move in the USA to increase the surveillance of adverse events associated with dietary supplements. Some of the pressure is coming from different ends of the spectrum: both advocates and opponents of these supplements. The advocates want a harmonized way to report these data. They don't necessarily want to report these data to the FDA, but they would like to see some organization collate these in a systematic fashion. They are confident that this would show that there are no major problems with dietary supplements, except in a few rare cases.

Przyrembel: The most important adverse effect we have identified recently is the use of nicotinic acid in relatively high doses. We are trying to get the manufacturers not to use it, but rather use nicotinamide. They still use nicotinic acid with doses in the order of 50 mg/day. There are enthusiastic consumers who take three a day, and adverse effects have been reported, not just flushing: someone with pre-existing cardiovascular disease suffered a myocardial infarction presumably in relation to a vascular collapse.

Scott: We have done intervention trials for in excess of 26 weeks on cardiovascular patients and normal subjects. The amount of folic acid you need daily to normalize or reduce homocysteine to its lowest common denominator is 200 µg/day. We are seeing values in the USA of 400–800 µg/day in significant proportions of the population. It is not absolutely true to say that lowering homocysteine to its normal value would prevent everything that it might prevent. It is just as an indicator of lowering the biomarker. What it does illustrate is that supplements are complicated. The whole NTD idea came out of supplement intervention, and then it turned out that you couldn't use supplements because people didn't take them. Then the idea was that we should tweak the diet with a bit of folic acid. Then we went into the modelling and found that it wasn't that easy. Now what has happened is that there are high levels. If one is buying supplements in a bottle you know how much you are taking. But the way the food industry uses additives leaves room for huge differences in intakes. We are now starting to get wide ranging exposures, which makes for an interesting experiment in the USA with 300 million people.

Halliwell: I'd like to pick up on the issue of people knowing what they are buying. When people buy a multivitamin they think it is the RDA of everything. Some of them are, but others aren't (too little or, quite often, far too much). The other thing I've seen is mixed supplements (e.g. with flavonoids plant extracts are added): you have to read the small print to know what is in there.

Coates: What is on the label is not necessarily what is in the bottle. There is considerable variability. There are products that have very reliable content information, but there are others for which the reported content is way off.

Scott: My point is that with the addition of vitamins to various foods (that is, fortified foods) there is a passive increase of quite staggering proportions when you have several possible different sources. It's not just folic acid but also other micronutrients.

Halliwell: One example is vitamin C. People take half a gram or a gram a day in a tablet, then they drink orange juice fortified with vitamin C, and then they buy 'healthy' candy, each piece of which contains the RDA of vitamin C. Before you know it they are eating above 2 g per day of the vitamin.

Aggett: Another example is vitamin A. This need not necessarily come from supplements. Vitamin A is added to animal feeds and as a result people who consume liver are at risk of much higher vitamin A intakes than is commonly appreciated.

Russell: With the plethora of supplements that are on the market and with no analysis of what is going on with regard to bioavailability or safety, it seems as if the regulatory agencies are set up to fail.

Taylor: There are two answers. First, in terms of what Beth was talking about with regards to independently building the body of evidence and in turn taking regulatory action, the current resources allow only very limited opportunity for that. However, second, there are statutory reviews for certain types of claims or statements—some have for example a 45 day period. This does give the agency to opportunity to review the data, but it is extremely challenging to realistically assess such data in the short time allowed.

Russell: So nothing is being done.

Taylor: Resources are very limited, but in fairness these are complicated issues overall that need time to coalesce.

Azzi: A little story. A colleague of mine from the University of Berne bought a bottle of 'Omega 3 from natural origin' that his daughter requested. After a certain time, she developed a serious mercury intoxication. The diagnosis took some time, but finally it was realised that the omega 3 fatty acid preparation contained a large amount of mercury. Is this possible?

Rayman: There are specific regulations to prevent this.

Przyrembel: There are maximum levels allowed for mercury in food. It is possible that the manufacturer is not a good one, and buys fish oil where he shouldn't.

Taylor: Heavy metal contamination takes resources to detect.

Przyrembel: All over the world control agencies have had cuts in their budget. Everyone is concerned with this

Ernst: Why did he buy it in the USA and not Switzerland?

Azzi: He was going to the USA and it was cheaper there.

Yetley: If it were a drug, the manufacturer would be required as part of GMPs to have tested for heavy metal toxicity. They would have records in their files which would be accessible to FDA auditors. If it is a supplement the FDA doesn't have access to manufacturer's data although the manufacturer still bears the burden to ensure that marketed products are not adulterated.

Rayman: In the UK our Food Standards Agency looked at fish oil supplements and found them all to be below the necessary levels for mercury.

Yetley: Several times at this meeting people have commented that we need to do some post-market surveillance and monitoring for potential problems. However, a word of caution is needed. For example, if you look at the β carotene/lung cancer trial, the increased risk of lung cancer in the treatment as compared to the placebo group was 18% (ATBC 1994). This level of increased risk would be unlikely to be detected in post-market surveillance and other types of observational studies. If you are relying on a post-marketing adverse event reporting system, acute severe responses may be detected but other types of effects (e.g. ones with a long latency, low incidence rates, commonly experienced symptoms with multiple causes) are less likely to be detected.

Halliwell: How do you define a catastrophe? It's something that's immediately obvious from an epidemiological study!

Boobis: Are the agencies not really concerned that we will have a whole slew of products of ill-defined specification available, and yet there will be no change to the approval process? The assumption that these will necessarily be safe is untenable. If you are doing a risk assessment of a product that requires approval and you have minimal information the answer is simple: it is not approved and the sponsor should go and obtain the data.

Yetley: The best offence is good data to show that this beginning assumption is not valid. We don't have that body of evidence yet.

Boobis: Are we going to get it? Nothing I have heard indicates that it will be forthcoming.

Aggett: I was struck by two comments that Beth made at the start, in terms of the involvement of scientists with the process. This is where the alarm signals can start to be raised and complacency can start to be challenged. In the light of your experience, is there a simple answer? Does one have to systematically engage scientific organizations as professional bodies? Do you have an extensive scientific advisory network?

Yetley: Government agencies often convene scientific advisory committees when appropriate. Or a government scientist may contact experts individually and informally for scientific advice. The biggest challenge is to obtain effective input from the scientific community when general public comment is solicited through notice and comment rulemaking. In this case, the scientific community tends not to get involved. However, communicating on a public policy issue may require different skills or approaches than communicating with a fellow scientist. Professional scientific organizations could help their members through training and educational programmes on how to make relevant and effective comments on public policy issues with a scientific component.

Taylor: There are two reasons why this happens. One is, as we have said, the nature of the authority the government has. But the other is resources. What is being discussed here is that there is the opportunity for scientists to communicate with their legislative bodies and give input as to getting good science for regulators and point to places where resources are needed for that purpose.

Aggett: Does congress have scientific advisory panels?

Taylor: Not specifically to my knowledge, although they may have staff members who are experts. But, of course, all of the US congressmen have constituents who talk to them, and what Beth is saying I think, is that not enough of the constituents who provide input are scientists or represent science organizations.

Aggett: This is worth systematic investigation in terms of policy development.

Yetley: There is a skillset needed. Professional scientific organizations could work to help their members. You have to be able to communicate differently with a politician than with a fellow scientist.

Katan: In the Netherlands we have a reasonable system for evaluating proposed interventions, namely the Netherlands Health Council. This is like the Institute of Medicine in the USA, and it is highly regarded by the government. The ministers tend to take their advice and put it into law. The problem with supplements is that they are in limbo, with governments saying that Brussels needs to regulate them and Brussels never getting anywhere. In that respect, Europe has suffered. Our national legislation on supplements would be much better if there were no EU. I believe we should say that we have tried to achieve a common EU policy on health claims for 25 years and it hasn't worked, so let's get together with those countries that take this seriously to make provisional laws.

Rayman: One thing I feel strongly about is the way the public is being duped by products that don't contain the ingredient they are claiming. Glucosamine is a good example: there are some glucosamine products on the market that have no glucosamine at all in them. If the regulatory authorities decided supplements had to have in them what they claimed, this would be of benefit to the public.

Halliwell: This is an issue of quality control.

Przyrembel: I think we shouldn't blame the regulatory agencies for quality control.

Aggett: In the UK this is a quality issue and would be dealt with through trading standards officers, and perhaps environmental health officers. Fortunately there is a link between some of the environmental health responsibilities with the food standards agency. At the moment the agency is looking at ways in which it can take a multi-agency approach to these sorts of issues. I have another question: did the Stevia episode lead to any revision of the regulatory process?

Yetley: No. It is allowed in supplements but not in conventional foods.

Coates: My main reason for being involved in this meeting was to get the best minds at work in helping us to identify the research issues that need to be developed here. My goal will be to see something that helps us to put the best science forward—science that can be used for many applications, including that of informing policy initiatives. One way that scientists can help is to make some recommendations about what the scientific agenda should look like.

Aggett: We need things that could stand up to rigorous external forensic scrutiny, when it comes to looking for the use of markers to assess intakes, exposures, outcomes and susceptibility. A lot of that we have talked about in terms of generating or examining hypotheses needs to be underpinned by sound science and practice. Beth Yetley, what approaches have been developed to address issues of intake?

Yetley: In the USA we have both supplement use and food intake data in the same databases. This is the best way to estimate total intakes from all sources. Secondly, we need more data like those gathered by the National Cancer Institute where they evaluated the validity of the intake data for energy and protein by checking reported intakes against results from measurements of doubly-labelled water and urinary nitrogen (Subar et al 2003). They found significant under-reporting biases. This type of validation of intake data is needed for other nutrients. At the Office of Dietary Supplements, we are developing two intake-related dietary supplement databases. First, we are developing a supplement database based on analytical results for selected nutrients in marketed products. Second, we are hoping to develop a database that contains label compositional information for marketed dietary supplement products.

Coates: There are databases that individual researchers and the CDC have developed to monitor dietary supplement intake in their populations. EPIC has such as database. The women's health initiative in the USA had a database. They are relatively incomplete and they are almost always purpose built to look at specific populations to assess intake. What has been tricky is that there are lots of variations, because there are lots of different kinds of products included in these. We would like to develop something that is comprehensive; more challenging is to create something that is dynamic, because this is a dynamic marketplace. Appropriate, relevant updates would be needed. This will take some doing.

Yetley: Another issue is to develop validated biomarkers of exposure.

Taylor: One thing that emerged from the FAO/WHO nutrient risk assessment workshop is that we need to harmonize the approach to collecting, handling and using data.

Aggett: At this workshop there was also some comment on agreed harmonization of approaches to intervention studies.

References

ATBC (Alpha-Tocopherol, Beta Carotene) Cancer Prevention Study Group 2004 The effect of vitamin E and beta carotene on the incidence of lung cancer and other cancers in male smokers. N Engl J Med 330:1029–1035

IOM (Institute of Medicine) 1989 Diet and Health: Implications for Reducing Chronic Disease. National Academy of Sciences Press, Washington, DC

Subar AF, Kipnis V, Trojano RP et al 2003 Using intake biomarkers to evaluate the extent of dietary misreporting in a large sample of adults: the OPEN study. Am J Epidmiol 158:1–13

Vitamin E

Roland Stocker[1]

Centre for Vascular Research, School of Medical Sciences, Faculty of Medicine, University of New South Wales, and Department of Haematology, Prince of Wales Hospital, UNSW Sydney, NSW 2052, Australia

Abstract. Vitamin E refers to a family of tocopherol and tocotrienol isomers discovered in 1922 as anti-infertility factor. Vitamin E deficiency causes infertility and delayed-onset ataxia in experimental animals, and it leads to neuronal dysfunctions in humans. However, based largely on its radical-scavenging antioxidant activity *in vitro*, vitamin E supplements are commonly thought to provide health benefits against diseases associated with oxidative damage, most notably cardiovascular diseases. Contrary to this belief, the outcome of recent large, prospective, randomized and placebo-controlled clinical studies does not encourage the use of vitamin E supplements. These overall disappointing results can be explained and substantiated by scientific data critically testing the strengths of evidence for many of the underlying assumptions and examining the possibility that *in vivo* vitamin E may have function(s) other than, or in addition to, acting as an antioxidant.

2007 Dietary supplements and health. Wiley, Chichester (Novartis Foundation Symposium 282) p 77–92

Background

Structure and definition of vitamin E

The naturally occurring compounds commonly classified as vitamin E include α-, β-, γ- and δ-tocopherols, and α-, β-, γ- and δ-tocotrienols (Fig. 1). Tocopherols and tocotrienols are substituted chromanols with a saturated (phytyl) and unsaturated side chain, respectively. Of these, only the 2*R*-stereoisomers of α-tocopherol, i.e. *RRR*-α-tocopherol[2] that occurs naturally in foods, and *RSR*-, *RRS*- and *RSS*-α-tocopherol contained in synthetic *all racemic (all rac)*-α-tocopherol, are retained in human plasma and tissues, and hence considered active vitamin E (RDI 2000). The 2*S*-stereoisomers of α-tocopherol in

---

[1] Current address: Centre for Vascular Research, Bosch Institute, Faculty of Medicine, University of Sydney, Medical Foundation Bldg K25, 92–94 Parramatta Road, Camperdown NSW 2006, Australia.
[2] Historically vitamin E is labeled as *d*-α-tocopherol or natural form (*i.e.*, *RRR*-α-tocopherol) or *dl*-α-tocopherol or *all rac*-α-tocopherol (*i.e. RRR*-, *RRS*-, *RSR*-, *RSS*-, *SSS*-, *SRS*-, *SSR*- and *SRR*-α-tocopherol).

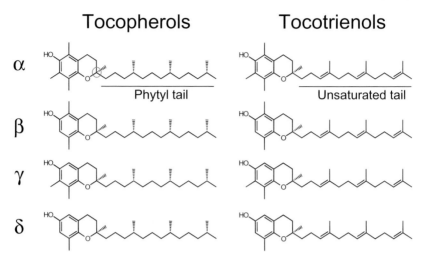

FIG. 1.   Naturally occurring forms of vitamin E. Structures classified conventionally as vitamin E include four isomers of each tocopherols and tocotrienols. Of these only the 2R-stereoisomeric forms of α-tocopherol (indicated by circle) are retained in the human body and hence considered to have vitamin E activity.

all *rac*-α-tocopherol used for fortified foods and vitamin supplements are not retained and thus not considered active. Considering the 2R-stereoisomers of α-tocopherol as active, *RRR*-α-tocopherol has twice the activity of *all rac*-α-tocopherol[3], so that 30 mg day of the *all rac*-α-tocopherol are required to achieve the recently recommended dietary allowance (RDA) of 15 mg/day of *RRR*-α-tocopherol (RDI 2000).

Function of vitamin E

Unlike other vitamins and most nutrients, a specific role for vitamin E in a required metabolic function has not been identified. Vitamin E was discovered as an anti-sterility factor (Evans & Bishop 1922). This led to the resorption–gestation test in rats as the functional assay for vitamin E activity, although its relevance to human biology remains unknown. Subsequent nutritional research realized that vitamin E was susceptible to destruction by dietary fats. In pioneering work, Mattill (1927) reported that a 'sterility' diet exposed to an oxygen atmosphere at 70 °C took up vastly more oxygen than a 'fertility' diet (i.e. sterility diet plus 2%

---

[3] According to official US Pharmacopcia (USP) conversions, one IU is defined as 1 mg of *all rac*-α-tocopheryl acetate. All conversions are based on rat fetal resorption assays.

wheat germ oil), and this was associated with development of a strong smell of rancidity. As a result, Mattill postulated a destructive action of the fat on the vitamin *outside* the body. This notion was gradually turned around to vitamin E, being oxidized itself, controlling the progress of oxidation in tissues. In this process, nutritional studies were succeeded by biochemical (Tappel & Zalkin 1960) and then chemical studies (Burton & Ingold 1981), and the link to fertility lost. Primarily as a result of these chemical studies, it is now well established that in organic solvents α-tocopherol is a highly effective chain-breaking antioxidant that interrupts the propagation step in free radical-mediated lipid peroxidation (Burton & Ingold 1986). Surprisingly however, clear proof for α-tocopherol inhibiting lipid peroxidation in humans is still not established. For example, using isoprostanes as a measure of *in vivo* lipid peroxidation, supplementation of healthy subjects with up to 2000 IU/day for up to 8 weeks did not significantly alter lipid peroxidation (Meagher et al 2001).

Absorption, metabolism and excretion

Intestinally absorbed vitamin E is secreted into mesenteric lymph chylomicrons that reach the circulation, where they are partially metabolized, providing some vitamin E for distribution to all circulating lipoproteins. The liver takes up the resulting chylomicron remnants with the remaining vitamin E, and secretes very low-density lipoproteins (LDL) enriched preferentially with α-tocopherol. This discrimination between tocopherols is achieved by hepatic α-tocopherol transfer protein (α-TTP) (Hosomi et al 1997), a critical determinant for the biological activity of vitamin E. Very low-density lipoproteins are metabolized to LDLs, during which some α-tocopherol is distributed to high-density lipoproteins, so that in a fasted person these two classes of lipoproteins each contain approximately half of the circulating α-tocopherol for distribution to non-hepatic tissues, with an apparent half-life of *RRR*-α-tocopherol in normal subjects of ~48 h. Faecal elimination is the major route of excretion of ingested vitamin E. Urinary excretion of the corresponding 2′carboxyethyl metabolites of tocopherols (*e.g.* α-CHEH) also takes place, particularly in situations of excess vitamin E intake (Brigelius-Flohé & Traber 1999).

Vitamin E adequacy

The prevention of hydrogen peroxide-induced haemolysis, a surrogate marker of early signs of vitamin E deficiency, is used to assess vitamin E adequacy (Bieri & Farrell 1976). Accordingly, plasma α-tocopherol concentrations of 12–14 μmol/l are adequate. On this basis, the RDA for vitamin E was set at 15 mg α-tocopherol per day for adults, men and women aged ≥21 years. A problem with the haemolysis test is that its clinical relevance is unclear. Vitamin E deficiency is so rare in humans that data on the relationship between dietary intakes of vitamin E and clinically relevant signs of vitamin E deficiency are not available. Indicators superior to the hydrogen peroxide-induced haemolysis test are presently not available. For example, the correlation between intake and plasma concentration of vitamin E greater than $16\,\mu mol/l$ is not strong, general markers of lipid peroxidation (e.g. isoprostanes) are considered to be too non-specific, and there is insufficient information available on the meaning of urinary α-CHEH levels or vitamin E biokinetics (RDI 2000).

Clinical effects of vitamin E deficiency

Vitamin E deficiency occurs only rarely and as a result of protein/energy malnutrition, fat malabsorption syndromes or genetic abnormalities in α-TTP. The primary clinical sign of a deficiency is a peripheral neuropathy, followed by spinocerebellar ataxia, skeletal myopathy, and retinopathy (Sokol 1993). Symptoms associated with vitamin E deficiency due to α-TTP defects and malabsorption syndromes can be reversed by vitamin E supplements if these are provided before irreversible damage has occurred.

Safety of vitamin E

There have been no reports on adverse effects associated with dietary vitamin E. The highest level of daily intake that is likely to pose no risk of adverse health effect (i.e. UL) has been set at 14 mg of any form of α-tocopherol per kg (RDI 2000). This level, determined on the basis of haemorrhagic effects in rats, corresponds to ~1000 mg α-tocopherol per day for an average weight person. Vitamin E is not mutagenic, carcinogenic or teratogenic in animals, and few side effects have been reported in humans with short-term supplements at $\leq 2100\,mg/d$ *RRR*-α-tocopherol.

In considering safety, one must recognize that randomized controlled trials (RCTs) on health benefits of dietary supplements are not designed to examine toxicity of a supplement and, in the case of vitamin E, there are some concerns. Two studies (Tardif et al 1997, Brown et al 2001) reported antioxidant cocktails containing vitamin E to blunt the efficacy of cardiovascular drugs, for reasons presently not understood. Also, a significant increase in heart failure and hospitalization for heart failure associated with 400 IU/d vitamin E supplements for 7 years has been reported (Lonn et al 2005). In that study, and in the primary prevention Women Health Study that randomly assigned 39 876 healthy women to 600 IU vitamin E or placebo for 10 years (Lee et al 2005), total mortality was slightly, although non-significantly, increased with vitamin E supplements (relative risk in both studies 1.04). In light of these studies and because studies on health benefit of α-tocopherol against neurodegenerative disorders have used vitamin E at up to 2000 IU/d, a careful meta-analysis of risk is warranted.

Relationship of vitamin E intake to disease

A number of chronic and degenerative diseases are associated with oxidative stress (Halliwell & Gutteridge 1999) and given its reported antioxidant activity, vitamin E supplements are believed by some to help prevent disease or even be useful as a medicine in secondary prevention of diseases. Contrary to this belief, however, the *totality* of currently available evidence from RCTs does not justify the use of vitamin E supplements to reduce the risk for cardiovascular disease (CVD), cancer or neurodegenerative diseases. The following focuses on the relationship of vitamin E intake to CVD, as this has been studied most extensively and the rationale for a beneficial effect of vitamin E is, arguably, stronger for this than other diseases.

Vitamin E and CVD

The idea that vitamin E may protect against CVD is based on the oxidative modification theory of atherosclerosis proposed originally in 1989 (Steinberg et al 1989). According to this theory, oxidized LDLs play a central role in the initiation and progression of atherosclerotic disease. Vitamin E was thought to be protective because it is the most abundant radical scavenger in LDLs and it was reported to attenuate other processes that promote atherosclerosis, i.e. endothelial dysfunction, leukocyte adhesion to endothelial cells, proliferation of smooth muscle cells, expression of scavenger receptor CD36 and platelet aggregation (Stocker & Keaney 2004). Furthermore, observational studies consistently found diets rich in vitamin E to be associated with decreased risk of CVD. However, large prospective, RCTs on vitamin E alone or in combination with other vitamins (e.g. vitamin C) or minerals (e.g. selenium) reveal no benefit on major CVD outcome (Table 1) (for primary references and more detail see Stocker & Keaney 2004, Pham & Plakogiannis 2005a). Vitamin E supplements were used at up to 600 IU/day and for up to 10 years.

Inherently, the interpretation of data from RCTs is complicated, and the overall conclusion that vitamin E supplements do not provide CVD benefit has been subjected to scrutiny (Jialal & Devaraj 2003). However, of the concerns raised, most appear increasingly difficult to sustain. For example, early encouraging results (Stephens et al 1996, Salonen et al 2003) could not be confirmed in larger RCTs (Yusuf et al 2000, Hodis et al 2002), even high dose of vitamin E administered for periods of up to 10 years (Lee et al 2005, Lonn et al 2005) were ineffective in large populations, and benefits reported for small ($n = 196$) and

TABLE 1   Randomized trials of vitamin E ± other antioxidants in the treatment and prevention of CVD

Trial	Subjects		CVD deaths		Non-fatal MI		Non-fatal stroke	
	Antiox(s)	Placebo	Antiox(s)	Placebo	Antiox(s)	Placebo	Antiox(s)	Placebo
Linxian (P)	14,792	14,792	249	274	—	—	—	—
ATBC (P)	14,564	14,569	853	870	—	—	—	—
CHAOS (S)	1,035	967	27	23	14	41	—	—
GISSI-P (S)	5,666	5,668	310	329	226	204	83	95
HOPE (S)	4,761	4,780	342	328	532	524	209	180
PPP (P)	2,231	2,264	22	26	19	18	20	13
HPS (S)	10,269	10,267	880	839	464	467	430	435
SUVIMAX (P)	6,364	6,377	—	—	134	137	—	—
WHS (P)	19,937	19,939	106	140	184	181	220	222
Total	**79,619**	**79,623**	**2,789**	**2,829**	**1,573**	**1,572**	**962**	**945**

Primary (P) and secondary (S) intervention trials with vitamin E alone (normal) or vitamin E plus other antioxidants (italic) and employing >1000 subjects are listed.

selected populations, such as haemodialysis patients with CVD (Boaz et al 2000), are questioned by lack of benefit in comparatively larger ($n = 993$) subgroups of large studies, such as men with mild–moderate renal insufficiency (Lonn et al 2005). Also, the common critique that supplement efficacy was not assessed in RCTs must be seen in the light that there is no established *in vivo* function of vitamin E (see above) and, where tested for large numbers of subjects, vitamin E at doses up to 2000 IU/d for up to 8 weeks does not inhibit *in vivo* lipid oxidation, as judged by urinary 4-hydroxynonenal and isoprostanes (Meagher et al 2001). Importantly, the types of RCT employed with vitamin E supplements have been successful in establishing the benefit of CVD drugs.

Vitamin E and experimental atherosclerosis

An important complication in the interpretation of data from RCTs on CVD outcome is the unknown relationship between CVD events and their primary underlying cause, i.e. atherosclerosis, that vitamin E is supposed to inhibit. However, and not commonly appreciated, the totality of evidence also does not support the notion that vitamin E supplements inhibit experimental atherosclerosis (Stocker & Keaney 2004) (Fig. 2). Thus, of 44 reported studies in the literature, 28 showed no significant benefit, five reported worsening of disease, and only seven showed a benefit that could be ascribed to vitamin E. The overall outcome looks different however for vitamin E-deficient animals, where four out of five studies reported a benefit for vitamin E supplements (Stocker & Keaney 2004). Also, vitamin E deficiency resulting from knocking out α-TTP, increases

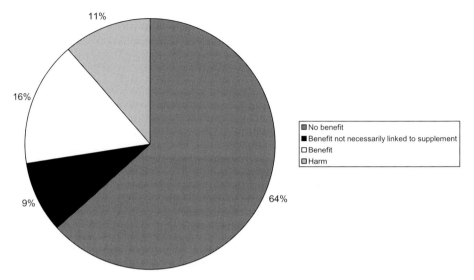

FIG. 2.   Effect of vitamin E supplements on experimental atherosclerosis. Of the 49 studies published until April 2004 and summarized here, 63.6% reported no benefit, 9.1% had benefit though this could not be linked to a direct action of the vitamin as supplements caused significant lipid-lowering, 15.9 % showed benefit and 11.4 % showed harm. For more details, see Stocker & Keaney (2004).

atherosclerosis in apolipoprotein E-null mice, albeit to a modest extent (10–30%) (Terasawa et al 2000). We recently confirmed these results, and showed that vitamin E supplements reversed this increase in disease without affecting lipid oxidation in the vessel wall (Suarna et al 2006).

Potential explanations for lack of benefit with vitamin E on CVD

The prediction that vitamin E supplements protect against atherosclerosis assumes that: there is a local 'deficiency' in the vitamin as disease progresses; LDL oxidation is an early event that initiates disease; vitamin E inhibits LDL oxidation in the vessel wall; and LDL oxidation is causally related to atherogenesis. Surprisingly, however, there is little direct evidence in support of these assumptions, rather, there is evidence to the contrary (Stocker & Keaney 2004). Briefly, even at the most advanced stages of atherosclerosis, lesions and their lipoproteins contain essentially unchanged concentrations of α-tocopherol, and there is no evidence for substantial vitamin E oxidation. Also, significant accumulation of lipoprotein lipid oxidation products follows (rather than precedes) the accumulation of non-oxidized lipids. In addition, these oxidized lipids appear to be formed in the presence

TABLE 2   **Relationship between lipid oxidation in diseased arteries and the extent of atherosclerosis**

Model	Treatment	Lipid oxidation	Disease	Reference
$Apoe^{-/-}Ldlr^{-/-}$ mice	Probucol bisphenol	↓	↓	(Witting et al 1999b)
WHHL rabbits	Probucol bisphenol	↓	→	(Witting et al 1999a)
$Apoe^{-/-}$ mice	Probucol	↓	→	(Witting et al 2000b)
$Apoe^{-/-}$ mice	Coenzyme Q_{10}	↓	↓	(Witting et al 2000a)
$Apoe^{-/-}$ mice	Probucol	↓[a]	↑[a]	(Choy et al 2005)
		↑[b]	↓[b]	
$Apoe^{-/-}$ mice	Red wine	→	↓	(Stocker & O'Halloran 2004)
$Apoe^{-/-}Ttpa^{-/-}$ mice	Vitamin E	→	↓	(Suarna et al 2006)

Lipid oxidation and atherosclerosis were assessed in the same arterial tissue.
[a] Aortic sinus.
[b] Thoracic and abdominal aorta.
$Apoe^{-/-}$, apolipoprotein E-deficient; $Apoe^{-/-}Ttpa^{-/-}$, apolipoprotein E- and α-tocopherol transfer protein-deficient; WHHL, Watanabe inheritable hyperlipidemic.

of α-tocopherol, an observation readily explained on mechanistic grounds related to the molecular action of the vitamin E in lipoproteins. Furthermore, at least in experimental animals, there is good evidence that lipoprotein lipid oxidation can be dissociated from (Table 2), and hence is not a cause for, atherosclerosis (Wu et al 2006).

The above findings provide a simple explanation for the lack of benefit of vitamin E supplements on atherosclerotic vascular disease and its outcome. They also challenge the commonly held view that vitamin E acts as an important, or even the most important, lipid-soluble antioxidant in humans (Burton et al 1982). As such, these observations may also help explain why vitamin E supplements do not provide clear protection against other oxidative stress-associated diseases, such as cancer, neurodegenerative diseases and cataract (Hanley & Miller 2005).

Vitamin E and cancer prevention and neurodegenerative disorders

Oxidative DNA damage is thought to contribute to carcinogenesis, so that vitamin E may inhibit the formation of certain cancers by either directly scavenging oxygen radicals that otherwise cause DNA damage, or by inhibiting lipid peroxidation that contributes to DNA oxidation. Similarly, vitamin E supplements could conceivably be beneficial against some neurodegenerative disorders that are associated with lipid and protein oxidation. However, similar to the situation in CVD, there is overall little evidence supporting such perceived benefit of vitamin E supplements (Pham & Plakogiannis 2005a, b).

Future research

For the last 50 years, most vitamin E research has focused on antioxidant activities, particularly on the ability of α-tocopherol to inhibit lipid oxidation, yet direct and strong support for the notion that vitamin E is an important inhibitor of *in vivo* lipid oxidation remains limited. Just as disappointingly, the specific biological role of vitamin E remains unclear more than 80 years after its discovery. Without doubt, vitamin E deficiency causes infertility in rodents that can be overcome by vitamin E supplementation, yet even for this benefit clear evidence for a contributing role of antioxidant activity remains speculation. Thus, it appears prudent to shift our research focus to non-antioxidant functions of vitamin E, a position taken for some years by Professor Angelo Azzi.

An obvious starting point is investigating how vitamin E deficiency causes abnormalities of syncytiotrophoblasts in the labyrinthine region in the placentas of living embryos, and how maternal blood-derived α-tocopherol protects these syncytiotrophoblasts to guarantee normal fetal viability and growth (Jishage et al 2005). It appears plausible that fetal up-take of α-tocopherol from maternal blood lipoproteins and transport of α-tocopherol within and between fetal cells is essential for its biological role, as mice deficient in transporters known to 'shuttle' α-tocopherol into (Rigotti et al 2003) and out of cells (Christiansen-Weber et al 2000, Oram et al 2001) are infertile. This may allow participation of α-tocopherol in transcriptional regulation of cellular processes relevant to placentation. It will be important to relate findings in the mouse to the human situation, given the marked differences in placental structure. A second area to redirect vitamin E research focus relates to the mechanisms by which vitamin E deficiency and proficiency affect the normal brain function related to neuropathy and ataxia.

Acknowledgements

The work of Roland Stocker has been supported from grants of the National Health & Medical Research Council of Australia and the Australian National Heart Foundation.

References

Bieri JG, Farrell PM 1976 Vitamin E. Vitam Horm 34:31–75
Boaz M, Smetana S, Weinstein T et al 2000 Secondary prevention with antioxidants of cardiovascular disease in endstage renal disease (SPACE): randomised placebo-controlled trial. Lancet 356:1213–1218
Brigelius-Flohé R, Traber MG 1999 Vitamin E: function and metabolism. FASEB J 13:1145–1155
Brown BG, Zhao XQ, Chait A et al 2001 Simvastatin and niacin, antioxidant vitamins, or the combination for the prevention of coronary disease. N Engl J Med 345:1583–1592

Burton GW, Ingold KU 1981 Autoxidation of biological molecules. 1. The antioxidant activity of vitamin E and related chain-breaking phenolic antioxidants in vitro. J Am Chem Soc 103:6472–6477

Burton GW, Ingold KU 1986 Vitamin E: Application of the principles of physical organic chemistry to the exploration of its structure and function. Acc Chem Res 19:194–201

Burton GW, Joyce A, Ingold KU 1982 First proof that vitamin E is major lipid-soluble, chain-breaking antioxidnat in human blood plasma. Lancet ii:327

Choy K, Beck K, Png FY et al 2005 Processes involved in the site-specific effect of probucol on atherosclerosis in apolipoprotein E gene knockout mice. Arterioscl Thromb Vasc Biol 25:1684–1690

Christiansen-Weber TA, Voland JR, Wu Y et al 2000 Functional loss of ABCA1 in mice causes severe placental malformation, aberrant lipid distribution, and kidney glomerulonephritis as well as high-density lipoprotein cholesterol deficiency. Am J Pathol 157:1017–1029

Evans HM, Bishop KS 1922 On the existence of a hitherto unrecognized dietary factor essential for reproduction. Science 56:650–651

Halliwell B, Gutteridge JMC 1999 Free radicals in biology and medicine. New York, Oxford University Press

Hanley DF, Miller ER 2005 An editorial update: *annus horribilis* for vitamin E. Ann Intern Med 143:143–145

Hodis HN, Mack WJ, LaBree L et al 2002 Alpha-tocopherol supplementation in healthy individuals reduces low- density lipoprotein oxidation but not atherosclerosis: the Vitamin E Atherosclerosis Prevention Study (VEAPS). Circulation 106:1453–1459

Hosomi A, Arita M, Sato Y et al 1997 Affinity for α-tocopherol transfer protein as a determinant of the biological activities of vitamin E analogs. FEBS Lett 409:105–108

Jialal I, Devaraj S 2003 Antioxidants and atherosclerosis: don't throw out the baby with the bath water. Circulation 107:926–928

Jishage K-I, Tachibe T, Ito T et al 2005 Vitamin E is essential for mouse placentation but not for embryonic development itself. Biol Reprod 73:983–987

Lee IM, Cook NR, Gaziano JM et al 2005 Vitamin E in the primary prevention of cardiovascular disease and cancer: the Women's Health Study: a randomized controlled trial. JAMA 294:56–65

Lonn E, Bosch J, Yusuf S et al 2005 Effects of long-term vitamin E supplementation on cardiovascular events and cancer. A randomized controlled trial. JAMA 293:1338–1347

Mattill HA 1927 The oxidative destruction of vitamins A and E and the protective action of certain vegetable oils. J Am Med Assoc 89:1505–1508

Meagher EA, Barry OP, Lawson JA, Rokach J, FitzGerald G 2001 Effects of vitamin E on lipid peroxidation in healthy persons. JAMA 285:1178–1182

Oram JF, Vaughan AM, Stocker R 2001 ATP-binding cassette transporter A1 mediates cellular secretion of α-tocopherol. J Biol Chem 276:39898–39902

Pham DQ, Plakogiannis R 2005a Vitamin E supplementation in cardiovascular disease and cancer prevention: Part 1. Ann Pharmacother 39:1870–1878

Pham DQ, Plakogiannis R 2005b Vitamin E supplementation in Alzheimer's disease, Parkinson's disease, tardive dyskinesia, and cataract: Part 2. Ann Pharmacother 39:2065–2072

RDI 2000 Vitamin E. RDI Dietary reference intakes for vitamin C, vitamin E, selenium, and carotenoids. The National Academy of Sciences, p 186–283

Rigotti A, Miettinen HE, Krieger M 2003 The role of the high-density lipoprotein receptor SR-BI in the lipid metabolism of endocrine and other tissues. Endocr Rev 24:357–387

Salonen RM, Nyyssönen K, Kaikkonen J et al 2003 Six-year effect of combined vitamin C and E supplementation on atherosclerotic progression: the Antioxidant Supplementation in Atherosclerosis Prevention (ASAP) Study. Circulation 107:947–953

Sokol RJ 1993 Vitamin E deficiency and neurological disorders. In: Packer L, Fuchs J (eds) Vitamin E in health and disease. Marcel Dekker, New York p 815–849

Steinberg D, Parthasarathy S, Carew TE, Khoo JC, Witztum JL 1989 Beyond cholesterol: Modifications of low-density lipoprotein that increase its atherogenicity. N Engl J Med 320:915–924

Stephens NG, Parsons A, Schofield PM et al 1996 Randomised controlled trial of vitamin E in patients with coronary disease: Cambridge Heart Antioxidant Study (CHAOS). Lancet 347:781–786

Stocker R, Keaney JF Jr 2004 Role of oxidative modifications in atherosclerosis. Physiol Rev 84:1381–1478

Stocker R, O'Halloran RA 2004 Dealcoholized red wine decreases atherosclerosis in apolipo-protein E gene-deficient mice independently of inhibition of lipid peroxidation in the artery wall. Am J Clin Nutr 79:123–130

Suarna C, Wu BJ, Choy K et al 2006 Protective effect of vitamin E supplements on experimental atherosclerosis is modest and depends on pre-existing vitamin E deficiency. Free Radic Biol Med 41:722–730

Tappel AL, Zalkin H 1960 Inhibition of lipid peroxidation in microsomes by vitamin E. Nature 185:35

Tardif J-C, Côté G, Lespérance J et al 1997 Probucol and multivitamins in the prevention of restenosis after coronary angioplasty. N Engl J Med 337:365–372

Terasawa Y, Ladha Z, Leonard SW et al 2000 Increased atherosclerosis in hyperlipidemic mice deficient in α-tocopherol transfer protein and vitamin E. Proc Natl Acad Sci USA 97:13830–13834

Witting PK, Pettersson K, Östlund-Lindqvist AM, Westerlund C, Wågberg M, Stocker R 1999a Dissociation of atherogenesis from aortic accumulation of lipid hydro(pero)xides in Wata-nabe heritable hyperlipidemic rabbits. J Clin Invest 104:213–220

Witting PK, Pettersson K, Östlund-Lindqvist AM, Westerlund C, Eriksson AW, Stocker R 1999b Inhibition by a co-antioxidant of aortic lipoprotein lipid peroxidation and atheroscle-rosis in apolipoprotein E and low density lipoprotein receptor gene double knockout mice. FASEB J 13:667–675

Witting PK, Pettersson K, Letters J, Stocker R 2000a Anti-atherogenic effect of coenzyme Q_{10} in apolipoprotein E gene knockout mice. Free Radic Biol Med 29:295–305

Witting PK, Pettersson K, Letters J, Stocker R 2000b Site-specific anti-atherogenic effect of probucol in apolipoprotein E deficient mice. Arterioscler Thromb Vasc Biol 20: e26–33

Wu BJ, Kathir K, Witting PK et al 2006 Antioxidants protect from atherosclerosis by a heme oxygenase-1 pathway that is independent of free radical scavenging. J Exp Med 203: 1117–1127

Yusuf S, Dagenais G, Pogue J, Bosch J, Sleight P 2000 Vitamin E supplementation and cardio-vascular events in high-risk patients. The heart outcomes prevention evaluation study inves-tigators. N Engl J Med 342:154–160

DISCUSSION

Halliwell: Vitamin E has been one of the most extensively studied vitamins, but from a nutritionist's point of view, are you now saying that supplements of it are a waste of time and have no biological effect? How much do we need in the diet?

Stocker: Vitamin E deficiencies are extremely rare, and only occur under the conditions that I listed. This suggests that the dietary intake normal people have is sufficient to allow for our bodies' requirements for vitamin E. Once we know more about its true biological function, I hope we'll be in a position to have better tools to reassess requirements. This may then lead to changes in recommendations.

Coates: How does excess vitamin E overcome the α-TTP deficiencies? Is it mass action?

Stocker: It is not really clear, though mass action is a likely explanation. People have tried to assess whether this is due to an antioxidant role by α-tocopherol. In the α-TTP-deficient mouse you can overcome sterility by supplementing animals with α-tocopherol or a synthetic analogue of it, BO-653 (Jishage et al 2001). The argument was made that this proves that it is an antioxidant effect of α-tocopherol *in vivo*. But if we look closely we see that in this study BO-653 was supplemented at 1000-times the concentration of α-tocopherol to overcome the infertility. Given that the bioavailability of BO-653 is ~10 times greater than that of α-tocopherol, and both compounds have comparable antioxidant activity, an antioxidant function of the vitamin in overcoming sterility does not appear likely.

Aggett: Your research points were very pertinent, but in some ways your presentation didn't lead up to them. Can you tell us more about the evidence for differential distribution of vitamin E? I understand that there is much to suggest that this is systemically controlled and responsive to oxidative stress (Blatt et al 2001).

Stocker: I don't recall the study you are referring to. When we think about vitamin E bioavailability, we must remember that this is a lipophilic compound. What I am about to say is relevant not only for vitamin E, but also for other lipid-soluble compounds, and one we have an interest in is coenzyme Q10. In fasting blood, about 50% of α-tocopherol is present in high-density lipoprotein (HDL), and the rest is in low-density lipoprotein (LDL). With regards to availability of α-tocopherol to brain, HDL (rather than LDL) is thought to be the major player (Balazs et al 2004). Vitamin E enters brain via a process referred to as selective uptake where lipid components of HDL are taken up without particle uptake. This 'selective uptake' is different to the process of lipid donation by LDL where entire lipoprotein particles are taken up. Brain is difficult to enrich with lipophilic substances, and if α-tocopherol is consumed in brain it will be more difficult to replenish it.

Aggett: Related to that, when measuring tocopherol levels in lesions, would you also get information by measuring it in adjacent areas that have not developed atherosclerosis?

Stocker: We have looked at both lesion tissue and adjacent non-lesion tissue, and find essentially little difference.

Manach: You mentioned drug–drug interactions. Do you think that beneficial interactions could also occur with micronutrients present in food?

Stocker: My comments related specifically to clinical trials. In those trials the differences were in the supplements and the drugs. The underlying assumption was that there were no differences in micronutrients. Two studies (Tardif et al 1997, Cheung et al 2001) reported an adverse effect of vitamin E in combination with other antioxidants on the efficacy of the drugs employed (probucol and simvastatin-niacin, respectively). Because vitamin E was used in combination with other antioxidants, these studies cannot reveal whether vitamin E alone was responsible for the adverse effects.

Azzi: I appreciate that the non-antioxidant function of vitamin E that I proposed more than 10 years ago has now been accepted (Boscoboinic et al 1991). There are other points to be discussed, especially regarding the prospective aspects that Roland has been pointing to, such as the gene expression analysis. These studies are not for the future. Already about a dozen of them (after our original work in this field; Ricciarelli et al 2000, 2001) are being carried out, showing that vitamin E produces increased or decreased expression of a number of genes; no compensatory antioxidant genes, however, are expressed in animals devoid of vitamin E (Gohil et al 2004, Barella et al 2004) as would be expected if vitamin E were acting as an antioxidant. There is also the finding that tocopherol can be phosphorylated, becoming tocopherol phosphate (Munteanu et al 2004, Negis et al 2005). This may be the active form of tocopherol *in vivo*, suggesting that tocopherol as such may not play an important role. I am concerned about the total negativity regarding the human studies on vitamin E. The small trials have to be considered very important and very often have given positive results. All the big trials have been dealing with highly ill individuals with coronary disease and calcified, occluded coronary arteries; the expectation that vitamin E would have any benefit in these patients is naïve. We need to go more in the direction of primary intervention studies. Lee et al (2005) showed that in women aged 50–65 taking a vitamin E supplement, there was a significant 24% reduction of cardiovascular death. Finally, the results of Meydani et al (1997) on the immune response stimulation by vitamin E in the elderly are important. In aged animals vitamin E is associated with a precise molecular event that stimulates the immune response. Older people who take vitamin E are also protected against infections of the upper respiratory tract.

Halliwell: How much vitamin E should we take in the diet?

Azzi: It is clear that if one wants to prevent the most obvious deficiency symptom, ataxia, the amount indicated by the FDA, 15 mg/day, is sufficient. We don't know what other endpoints we should be looking at that require higher vitamin E intake to avoid pathological events. Even the most pessimistic investigations, such as that made in the meta-analysis by Edgar Miller (Miller et al 2005), 200 mg/day of vitamin E produces no damage. I would recommend an amount between 15 mg and 200 mg daily.

Stocker: The fact is that vitamin E was discovered 1922, yet we still don't know its function. As a scientist, I don't think this is a strong track record. With regards to dietary intake, I think the panel responsible for FDA's 15 mg/day recommendation overall did a good job. Unless we have more precise knowledge about the function of vitamin E in the body, I don't think we have a basis to come up with a substantially different recommendation.

Rayman: Practically all the research we have looked at is on α-tocopherol. The major form of tocopherol in the US diet is γ-tocopherol. There is some evidence that γ-tocopherol could be a much more important player in cancer prevention than α-tocopherol.

Azzi: That is a good point. Another element which led us to disprove the concept that vitamin E is an antioxidant is that each one of the different vitamin E analogues has its own effects *in vitro* and *in vivo* although all of them have similar antioxidant properties. γ-Tocopherol is important because it is present in the American diet. It appears to specifically protect against prostate carcinoma. Roland Stocker stated that there is no molecular evidence for what tocopherols are doing. There is actually strong evidence, coming from a number of laboratories, that tocopherols are gene regulators (Barella et al 2004, Azzi et al 1998, Gohil et al 2003). There is also evidence that there is a particular element in all the genes that are regulated by vitamin E, which could be considered to be a 'vitamin E-responsive element', responsible for gene expression control.

Russell: Are vitamin–vitamin interactions or vitamin–drug interactions a concern with vitamin E? For example, vitamin E–vitamin K–warfarin interactions?

Stocker: There have been reports on potential interaction of warfarin with vitamin E (Heck et al 2000), and this may relate to vitamin K deficiency. However, the extent to which this occurs is limited.

Russell: Many antioxidants at high dose can actually act as prooxidants. Is there any evidence that at high tissue concentrations, vitamin E acts as a prooxidant?

Stocker: By definition, every antioxidant has the potential to be a prooxidant. Carotenoids are probably more likely to be oxidants than antioxidants. There is limited evidence that vitamin E acts as a prooxidant *in vivo*, although this is not an area that has been studied in detail.

Alexander: There is a more general problem here. When we discuss various vitamins we are actually bundling together a number of substances with different chemical structures. The international recommendations usually relates to only one property of these vitamins, namely those of nutrition. They might differ with respect to various biochemical mechanisms and regarding adverse effects. With vitamin E, one of the compounds has a high affinity to a transport protein, whereas others are linked to the nuclear receptor. It is important for this field to talk more about specific chemical compounds rather than lumping them together.

Azzi: I appreciate your comment. When we are talking about vitamin E we are talking about a diffuse concept. If these compounds are important, as modulators of signal transduction and gene expression, they can't be oxidised and reduced *ad libitum*; they have to be kept protected like hormones, such as oestrogens and androgens, to prevent their degradation. Retinal is another good example. It must be protected because we don't want to suddenly become blind when there is a little more oxygen around. In this sense, vitamin E is not protecting against free radicals but it has to be protected.

Shekelle: I want to comment from a different perspective. For 20 years I have been assessing the effect of interventions in humans. It is mostly trial evidence: the effect of an intervention applied to humans on an outcome. For the NCCAM we reviewed the vitamin E data about 5 years ago. There is no signal there. In all my work on interventions in humans, we usually either see a signal, and that signal is consistent across trials and across subgroups in trials, or you don't see a signal. When you don't see a signal there is usually nothing there. Roland Stocker made the point that other drugs tested in this same fashion do show signals, and these are consistent across all subgroups. This doesn't mean that vitamin E doesn't work (it might), but from where I sit, the burden of proof is now on the proponents of vitamin E to identify that compound, subgroup and outcome where it really does have an effect.

References

Azzi A, Boscoboinik D, Fazzio A et al 1998 RRR-alpha-tocopherol regulation of gene transcription in response to the cell oxidant status. Z Ernahrungswiss 37:21–28

Balazs Z, Panzenboeck U, Hammer A et al 2004 Uptake and transport of high-density lipoprotein (HDL) and HDL-associated α-tocopherol by an in vitro blood–brain barrier model. J Neurochem 89:939–950

Barella L, Muller PY, Schlachter M et al 2004 Identification of hepatic molecular mechanisms of action of alpha-tocopherol using global gene expression profile analysis in rats. Biochim Biophys Acta 1689:66–74

Blatt DH, Leonard SW, Traber MG 2001 Vitamin E kinetics and the function of tocopherol regulatory proteins. Nutrition 17:799–805

Boscoboinik D, Szewczyk A, Hensey C, Azzi A 1991 Inhibition of cell proliferation by alpha-tocopherol. Role of protein kinase C. J Biol Chem 266:6188–6194

Cheung MC, Zhao XQ, Chait A, Albers JJ, Brown BG 2001 Antioxidant supplements block the response of HDL to simvastatin-niacin therapy in patients with coronary artery disease and low HDL. Arterioscler Thromb Vasc Biol 21:1320–1326

Gohil K, Schock BC, Chakraborty AA et al 2003 Gene expression profile of oxidant stress and neurodegeneration in transgenic mice deficient in alpha-tocopherol transfer protein. Free Radic Biol Med 35:1343–1354

Gohil K, Godzdanker R, O'Roark E et al 2004 Alpha-tocopherol transfer protein deficiency in mice causes multi-organ deregulation of gene networks and behavioral deficits with age. Ann NY Acad Sci 1031:109–126

Heck AM, DeWitt BA, Lukes AL 2000 Potential interactions between alternative therapies and warfarin. Am J Health Syst Pharm 57:1221–1227

Jishage K, Arita M, Igarashi K et al 2001 Alpha-tocopherol transfer protein is important for the normal development of placental labyrinthine trophoblasts in mice. J Biol Chem 276:1669–1672

Lee IM, Cook NR, Gaziano JM et al 2005 Vitamin E in the primary prevention of cardiovascular disease and cancer: the Women's Health Study: a randomized controlled trial. JAMA 294:56–65

Meydani SN, Meydani M, Blumberg JB et al 1997 Vitamin E supplementation and in vivo immune response in healthy elderly subjects. A randomized controlled trial. JAMA. 277:1380–1386

Miller ER 3rd, Pastor-Barriuso R, Dalal D, Riemersma RA, Appel LJ, Guallar E 2005 Meta-analysis: high-dosage vitamin E supplementation may increase all-cause mortality. Ann Intern Med 142:37–46

Munteanu A, Zingg JM, Ogru E et al 2004 Modulation of cell proliferation and gene expression by alpha-tocopheryl phosphates: relevance to atherosclerosis and inflammation. Biochem Biophys Res Commun 318:311–316

Negis Y, Zingg J-M, Ogru E, Gianello R, Libinaki R, Azzi A 2005 On the existence of cellular tocopheryl phosphate, its synthesis, degradation and cellular roles: a hypothesis. IUBMB Life 57:23–25

Ricciarelli R, Zingg JM, Azzi A 2001 Vitamin E: protective role of a Janus molecule. FASEB J 15:2314–2325

Ricciarelli R, Zingg JM, Azzi A 2000 Vitamin E reduces the uptake of oxidized LDL by inhibiting CD36 scavenger receptor expression in cultured human aortic smooth muscle cells. Circulation 102:82–87

Tardif JC, Cote G, Lesperance J et al 1997 Probucol and multivitamins in the prevention of restenosis after coronary angioplasty. Multivitamins and Probucol Study Group. N Engl J Med 337:365–372

Flavonoids: a re-run of the carotenoids story?

Barry Halliwell

Department of Biochemistry, Yong Loo Lin School of Medicine, National University of Singapore, 8 Medical Drive, MD7 Level 2, Singapore 117597

Abstract. Flavonoids have powerful antioxidant activities *in vitro*, but the evidence that they act as antioxidants *in vivo* in humans is equivocal at best. However, they may be able to help protect the gastro-intestinal tract against reactive oxygen species.

2007 Dietary supplements and health. Wiley, Chichester (Novartis Foundation Symposium 282) p 93–104

Setting the scene

Oxygen is toxic, and we only survive its presence because we have evolved a plethora of antioxidant defence systems. These systems minimize the levels of oxygen radicals and other reactive oxygen species (ROS) but do not eliminate them completely, since some ROS are useful (Halliwell & Gutteridge 2007). Thus some ROS-dependent 'oxidative damage' occurs continually in the human body, measurable by various 'biomarkers' such as F_2-isoprostanes (products of oxidation of lipids) and oxidized DNA bases, such as 8-hydroxy-2′-deoxyguanosine (8OHdG) (Halliwell & Whiteman 2004). This oxidative damage is thought to contribute to the age-related development of cancer, cardiovascular and neurodegenerative diseases, and several other disorders, and perhaps even to the ageing process itself (Beckman & Ames 1997, Sohal et al 2002, Butterfield & Boyd-Kimball 2004, Halliwell & Gutteridge 2007). To some extent this may be a product of evolution—ROS are involved in a network of signalling processes that mount the body's response to infection, and they can aid killing of bacteria and viruses (Babior 2004). When humans first gathered together in cities, infectious disease was rampant, driving the evolution of powerful immune responses to which ROS contribute (Babior 2004, Fang 2004). Thus the ability to make a lot of ROS might be selected for, keeping young people alive to reproduce. It doesn't matter to evolution if ROS give you cancer or other diseases in your later (post-reproductive) years (Halliwell 2004).

The position of plants

Green plants have a special problem with O_2 toxicity, being exposed to the full force of the pure O_2 that they produce during photosynthesis. One therefore expects them to be loaded with antioxidants and indeed they are: vitamin C, carotenoids, tocopherols, tocotrienols and the multitudinous polyphenols, such as the flavonoids. All these molecules seem to be important antioxidants in plants, although of course they have other metabolic roles as well (Halliwell & Gutteridge 2007). Humans must obtain vitamins E and C, carotenoids and flavonoids from plants, since we cannot make them ourselves.

Epidemiological studies in the 1980s and 1990s revealed that humans with high intakes (or blood levels) of vitamin C, α-tocopherol, β-carotene and other carotenoids from their diet are less likely, on average, to suffer myocardial infarctions, other vascular disease, diabetes and many forms of cancer (Gey 1995). These studies coincided with intense research on the biological importance of oxygen radicals, other ROS (such as H_2O_2 and peroxynitrite) and antioxidant defences *in vivo*. It was discovered that increased oxidative damage accompanies most, if not all, human diseases and contributes to the pathology of several, e.g. cigarette smoke-induced lung cancer, chronic inflammation, atherosclerosis and Alzheimer's disease (Beckman & Ames 1997, Butterfield & Boyd-Kimball 2004, Halliwell & Gutteridge 2007). Putting two and two together, it was widely assumed that these antioxidants were protective agents—taking them in the diet or as supplements or in fortified foods should decrease oxidative damage and diminish disease incidence.

The gold standard of epidemiology is the double-blind placebo-controlled intervention trial. That's when it started to go wrong. The ATBC (α-tocopherol/β-carotene) study in Finland revealed that α-tocopherol supplements had no effect on lung cancer incidence in heavy smokers, but β-carotene supplements increased the risk (Virtamo et al 2003). Several other studies on various populations revealed little or no effect of vitamins C, E and β-carotene on disease prevention in well-nourished subjects, and a few suggestions of harm from high doses taken for long periods (Virtamo et al 2003, Bjelakovic et al 2004, Neuhouser et al 2004, Miller et al 2005, Lee et al 2004b, Blacker 2005, Lawlor et al 2004).

So how do we explain this? It is widely agreed by nutritionists that diets rich in vegetables and fruits are associated with lowered incidence of cardiovascular disease, dementia, diabetes, stroke and certain types of cancer, especially lung and oral cancers. The more vegetables and fruits you eat, the greater will be your body content of antioxidants. However, plants contain a huge range of agents that might protect against disease (reviewed in Halliwell 2006). In addition, a fruit- and vegetable-rich diet is often low in fat, and high fat intake is a risk factor for cardiovascular disease, diabetes and some cancers, and it promotes oxidative stress

(Morrow 2005). Thus it could be anything in the dietary plants that protects against disease, and high body antioxidant levels could be a 'biomarker' of a good diet. If so, reproducing these levels with supplements may not give the same benefit (Halliwell 1999).

Antioxidants and oxidative damage

Does the failure of vitamins E, C and β-carotene to protect against the development of age-related diseases mean that ROS are unimportant as contributors to disease pathology? Actually, no. Almost all studies assumed that feeding antioxidants would decrease oxidative damage without measuring such damage to prove that it did decrease. Yet we now know that these 'antioxidants' often do not decrease oxidative damage *in vivo*. For example, 60 mg daily of ascorbate seems sufficient to minimize oxidative DNA damage in humans, and more has no further effect (Halliwell & Gutteridge 2007). High-dose α-tocopherol is poorly-effective at decreasing levels of lipid peroxidation in healthy humans, when measured by reliable biomarkers (Meagher et al 2001). It is much better at decreasing lipid peroxidation in mice, and its anti-atherosclerotic effects are correspondingly greater (Pratico et al 1998). Studies in Denmark showed that urinary excretion of 8OHdG was decreased about 28% by feeding Brussels sprouts to volunteers (Verhagen et al 1997), but not by supplementing these subjects with β-carotene, vitamin C, or α-tocopherol (Prieme et al 1997). We found that a mixture of antioxidants could sometimes transiently increase oxidative DNA damage (Halliwell 1999). Indeed, plasma F_2-isoprostane levels respond better to weight loss, good diet and lowering plasma cholesterol levels than they do to antioxidant supplements (Morrow 2005, Meagher et al 2001). Overall, for healthy subjects, there seems to be little benefit (and possible harm) of consuming high-dose supplements of single antioxidants. Never consume β-carotene supplements if you smoke.

Does this mean that fruits and vegetables are beneficial for reasons other than their antioxidant content, or that the most important antioxidants in them have not yet been identified? Probably both are true. Plants contain multiple agents protective against disease that are not antioxidants. Indeed, some may be mild pro-oxidants, increasing the levels of endogenous defence systems by creating some degree of oxidative stress (Laughton et al 1991, Velayutham et al 2005). Several authors have shown that consumption of antioxidant-rich foods decreases levels of oxidative damage *in vivo* in humans (Lee et al 2006; reviewed by Halliwell et al 2005). Others have found little effect (e.g. McAnulty et al 2005), and a few studies registered increases in biomarkers of oxidative protein damage, such as 2-aminoadipic and γ-glutamyl semialdehydes (Dragsted et al 2004). One must be very cautious in such studies to rule out confounding effects of refeeding fasted individuals, as opposed to the effects of antioxidants in the food, on biomarkers of oxidative damage (Lee et al 2004a, 2006, Richelle et al 1999). Nevertheless, the bulk of evidence does suggest that antioxidant effects do contribute to the benefits of a high intake of fruits and vegetables (Halliwell et al 2005), although these effects cannot be reproduced by supplements of ascorbate, vitamin E, or β-carotene.

So now to the flavonoids

Flavonoids and other polyphenols have powerful antioxidant activities *in vitro*, being able to scavenge a wide range of ROS. Many can chelate transition metal ions such as iron and copper, decreasing their ability to promote oxidative damage *in vitro* (Rice-Evans 2000). Two observations drew attention to their potential importance. First, flavonoids in red wine were shown to be able to inhibit the oxidation of low density lipoproteins, and this was suggested as an explanation of the 'French paradox' (Frankel et al 1993). Second, the Zutphen study, an epidemiological study in the Netherlands, suggested an inverse correlation between the incidence of coronary heart disease and stroke and the dietary intake of flavonoids, especially quercetin (Hertog et al 1993). Since then multiple other epidemiological studies have confirmed similar associations, although a few have not, and there is little evidence of protection against cancer (Neuhouser 2003).

Thus could flavonoids be major contributors to the disease-protective effects of fruits and vegetables? Many polyphenols are absorbed, although rarely completely, and the remainder metabolized in the colon to generate high levels of monophenols (Manach & Donovan 2004, Jenner et al 2005). But are they better antioxidants than vitamins C, E and β-carotene *in vivo*? Again, studies with biomarkers have given a mixture of results, and a review by Halliwell et al (2005) concluded that the balance of evidence overall did not support significant systemic antioxidant effects of absorbed flavonoids. Indeed, since plasma levels of unconjugated flavonoids rarely exceed 1 μM and the metabolites tend to have lower antioxidant activity because of the blocking of radical-scavenging OH groups by methylation, sulfation or glucuronidation (Williamson et al 2005), it seems difficult to imagine a powerful antioxidant effect *in vivo*. Some studies have shown effects of flavonoid-rich foods in raising plasma total antioxidant capacity (TAC). But one must be cautious here; many such foods can increase plasma uric acid levels, and urate is detected by several TAC assays. Since elevated urate may be a risk factor for some diseases, the alleged 'antioxidant benefit' may not be what it seems (Halliwell 2003a, Lotito & Frei 2004). Finally, flavonoids and other phenols are complex molecules that have multiple actions *in vivo*, including inhibiting telomerase, affecting signal transduction pathways, inhibiting cyclooxygenases and lipoxygenases, decreasing xanthine oxidase, matrix metalloproteinase, angiotensin-converting enzyme, proteasome, and sulphotransferase activities, and interacting with sirtuins. Flavonoids may also interact with cellular drug transport systems, compete with glucose for transmembrane transport, interfere with cyclin-dependent regulation of the cell cycle, inhibit protein glycation, and affect platelet function (e.g. Howitz et al 2003, van Hoorn et al 2002, Laughton et al 1991, Spencer et al 2001, Naasani et al 2003). Again, it is uncertain whether these effects can happen systemically.

Aiding the gastro-intestinal (GI) tract

Halliwell et al (2000) proposed that antioxidant and other protective effects of flavonoids and other phenolic compounds could occur before absorption, within the stomach, intestines and colon. This could account for the suggested ability of flavonoid-rich foods to protect against gastric, and possibly colonic, cancer, although again it must not be assumed that any protective effect of flavonoid-rich foods is attributable to antioxidant actions of the flavonoids, or to flavonoids at all, rather than to other components in the foods. However, ingestion of green tea was reported to rapidly decrease prostaglandin E_2 concentrations in human rectal mucosa, consistent with inhibition of cyclooxygenase activity (August et al 1999), a potential anticancer mechanism.

The logic behind this hypothesis is that phenolic compounds present in plasma at $\leq 1\,\mu M$ concentrations are present in the stomach and intestines at much higher concentrations after consumption of foods and beverages rich in such compounds (Jenner et al 2005). Because absorption of phenolic compounds is incomplete, they enter the colon, where they and their products of bacterial metabolism can exert beneficial effects. Indeed, faecal water contains micromolar levels of flavonoids, and much higher levels of monophenols, and levels of flavonoids in the stomach and intestines will be even higher (Jenner et al 2005). Why should this be important? The gastro-intestinal (GI) tract is constantly exposed to ROS, both endogenously produced and from the diet. The stomach is especially affected by the latter; indeed, Kanner and Lapidot (2001) referred to the stomach as a 'bioreactor'. Sources of ROS include the mixtures of ascorbate and Fe^{2+} in the stomach (dietary iron, dietary ascorbate, and ascorbate normally present in gastric juice), haem proteins (also potential powerful pro-oxidants), lipid peroxides, cytotoxic aldehydes, and isoprostanes in the diet. Nitrite in saliva and in foods is converted to HNO_2 by gastric acid, forming nitrosating and DNA-deaminating species. There are also high concentrations of H_2O_2 in certain beverages, which can contain oxidizable, pro-oxidant, phenolic compounds such as hydroxyhydroquinone. Activation of immune cells naturally present in the GI tract by diet-derived bacteria and toxins can also increase ROS production (Halliwell & Gutteridge 2007).

Flavonoids and other phenolic compounds might exert direct protective effects in the GI tract, by scavenging ROS. They can inhibit haem protein-induced peroxidation in the stomach (Kanner & Lapidot 2001). They are able to inhibit DNA base deamination by HNO_2-derived reactive nitrogen species (Zhao et al 2001), up-regulate toxin-metabolizing or antioxidant defence enzymes in the GI tract and chelate transition metal ions to decrease their pro-oxidant potential (Halliwell et al 2005). Dietary iron is usually not completely absorbed, especially among subjects on Western diets. Unabsorbed dietary iron enters the faeces, where it could represent a pro-oxidant challenge to the colon and rectum (Babbs 1990). Indeed, diets rich in fat and low in fibre may aggravate this pro-oxidant effect. Phenolic compounds, by chelating iron, may help to alleviate pro-oxidant actions of colonic iron.

Artefacts of cell culture

Many studies examining the cytotoxic and other effects of flavonoids on malignant, and other, cells in culture may have been led astray by artefacts. Flavonoids oxidize readily in many commonly-used cell-culture media, generating H_2O_2, quinones and semiquinones that can contribute to cytotoxicity (Long et al 2000, Wee et al 2003, Halliwell 2003b). For example, the apparent cytotoxicity of green tea to PC12 cells was purely artefactual (Chai et al 2003). If flavonoids really are anticancer agents, more experiments demonstrating this *in vivo* are required.

Conclusion

To the author, flavonoids are typical xenobiotics, metabolized as such and rapidly removed from the circulation. High levels may even be toxic, but low levels of toxins can sometimes be good for you by raising levels of xenobiotic-metabolizing and antioxidant defence enzymes. Thus stick to flavonoid-rich foods; red wine (alcohol in moderation is good for you), tea, fruits and vegetables. Don't start taking high-dose supplements or foods heavily fortified with flavonoids until we know more; we do not want a repeat of the β-carotene error (Hercberg 2005). As I said a while ago, 'a protective effect of diet is not equivalent to a protective effect of antioxidants in diet' (Halliwell 2000).

References

August DA, Landau J, Caputo D, Hong J, Lee MJ, Yang CS 1999 Ingestion of green tea rapidly decreases prostaglandin E2 levels in rectal mucosa in humans. Cancer Epidemiol Biomarkers Prev 8:709–713
Babbs CF 1990 Free radicals and the etiology of colon cancer. Free Radic Biol Med 8:191–200
Babior BM 2004 NADPH oxidase. Curr Opin Immunol 16:42–47
Beckman KB, Ames BN 1997 Oxidative decay of DNA. J Biol Chem 272:19633–19636
Bjelakovic G, Nikolova D, Simonetti RG, Gluud C 2004 Antioxidant supplements for prevention of gastrointestinal cancers: a systematic review and meta-analysis. Lancet 364:1219–1228

Blacker D 2005 Mild cognitive impairment—no benefit from vitamin E, little from donepezil. N Engl J Med 352:2439–2441

Butterfield DA, Boyd-Kimball D 2004 Proteomics analysis in Alzheimer's disease: new insights into mechanisms of neurodegeneration. Int Rev Neurobiol 61:159–188

Chai PC, Long LH, Halliwell B 2003 Contribution of hydrogen peroxide to the cytotoxicity of green tea and red wines. Biochem Biophys Res Commun 304:650–654

Dragsted LO, Pedersen A, Hermetter A et al 2004 The 6-a-day study: effects of fruit and vegetables on markers of oxidative stress and antioxidative defense in healthy nonsmokers. Am J Clin Nutr 79:1060–1072

Fang FC 2004 Antimicrobial reactive oxygen and nitrogen species: concepts and controversies. Nat Rev Microbiol 2:820–832

Frankel EN, Kanner J, German JB, Parks E, Kinsella JE 1993 Inhibition of oxidation of human low-density lipoprotein by phenolic substances in red wine. Lancet 341:454–457

Gey KF 1995 Cardiovascular disease and vitamins. Concurrent correction of 'suboptimal' plasma antioxidant levels may, as important part of 'optimal' nutrition, help to prevent early stages of cardiovascular disease and cancer, respectively. Bibl Nutr Dieta 52:75–91

Halliwell B 1999 Establishing the significance and optimal intake of dietary antioxidants: the biomarker concept. Nutr Rev 57:104–113

Halliwell B 2000 The antioxidant paradox. Lancet 355:1179–1180

Halliwell B 2003a Plasma antioxidants: health benefits of eating chocolate? Nature 426:787; discussion 788

Halliwell B 2003b Oxidative stress in cell culture: an under-appreciated problem? FEBS Lett 540:3–6

Halliwell B 2004 Ageing and disease: from Darwinian medicine to antioxidants? Innovation 4:13–16

Halliwell B 2006 Polyphenols; antioxidant treats for healthy living or covert toxins? J Sci Food Agric 86:1992–1995

Halliwell B, Whiteman M 2004 Measuring reactive species and oxidative damage in vivo and in cell culture: how should you do it and what do the results mean? Br J Pharmacol 142:231–255

Halliwell B, Gutteridge JMC 2007 Free Radicals in Biology and Medicine, 4th edn. Oxford University Press

Halliwell B, Zhao K, Whiteman M 2000 The gastrointestinal tract: a major site of antioxidant action? Free Radic Res 33:819–830

Halliwell B, Rafter J, Jenner A 2005 Health promotion by flavonoids, tocopherols, tocotrienols, and other phenols: direct or indirect effects? Antioxidant or not? Am J Clin Nutr 81:268S–276S

Hercberg S 2005 The history of beta-carotene and cancers: from observational to intervention studies. What lessons can be drawn for future research on polyphenols? Am J Clin Nutr 81:218S–222

Hertog MG, Feskens EJ, Hollman PC, Katan MB, Kromhout D 1993 Dietary antioxidant flavonoids and risk of coronary heart disease: the Zutphen Elderly Study. Lancet 342:1007–1011

Howitz KT, Bitterman KJ, Cohen HY et al 2003 Small molecule activators of sirtuins extend Saccharomyces cerevisiae lifespan. Nature 425:191–196

Jenner AM, Rafter J, Halliwell B 2005 Human fecal water content of phenolics: the extent of colonic exposure to aromatic compounds. Free Radic Biol Med 38:763–772

Kanner J, Lapidot T 2001 The stomach as a bioreactor: dietary lipid peroxidation in the gastric fluid and the effects of plant-derived antioxidants. Free Radic Biol Med 31:1388–1395

Laughton MJ, Evans PJ, Moroney MA, Hoult JR, Halliwell B 1991 Inhibition of mammalian 5-lipoxygenase and cyclo-oxygenase by flavonoids and phenolic dietary additives.

Relationship to antioxidant activity and to iron ion-reducing ability. Biochem Pharmacol 42:1673–1681

Lawlor DA, Davey Smith G, Kundu D, Bruckdorfer KR, Ebrahim S 2004 Those confounded vitamins: what can we learn from the differences between observational versus randomised trial evidence? Lancet 363:1724–1727

Lee CY, Jenner AM, Halliwell B 2004a Rapid preparation of human urine and plasma samples for analysis of F2-isoprostanes by gas chromatography-mass spectrometry. Biochem Biophys Res Commun 320:696–702

Lee DH, Folsom AR, Harnack L, Halliwell B, Jacobs DR Jr 2004b Does supplemental vitamin C increase cardiovascular disease risk in women with diabetes? Am J Clin Nutr 80: 1194–1200

Lee CY, Isaac HB, Wang H et al 2006 Cautions in the use of biomarkers of oxidative damage; the vascular and antioxidant effects of dark soy sauce in humans. Biochem Biophys Res Commun 344:906–911

Long LH, Clement MV, Halliwell B 2000 Artifacts in cell culture: rapid generation of hydrogen peroxide on addition of (−)-epigallocatechin, (−)-epigallocatechin gallate, (+)-catechin, and quercetin to commonly used cell culture media. Biochem Biophys Res Commun 273:50–53

Lotito SB, Frei B 2004 The increase in human plasma antioxidant capacity after apple consumption is due to the metabolic effect of fructose on urate, not apple-derived antioxidant flavonoids. Free Radic Biol Med 37:251–258

Manach C, Donovan JL 2004 Pharmacokinetics and metabolism of dietary flavonoids in humans. Free Radic Res 38:771–785

McAnulty SR, McAnulty LS, Morrow JD et al 2005 Effect of daily fruit ingestion on angiotensin converting enzyme activity, blood pressure, and oxidative stress in chronic smokers. Free Radic Res 39:1241–1248

Meagher EA, Barry OP, Lawson JA, Rokach J, FitzGerald GA 2001 Effects of vitamin E on lipid peroxidation in healthy persons. JAMA 285:1178–1182

Miller ER 3rd, Pastor-Barriuso R, Dalal D, Riemersma RA, Appel LJ, Guallar E 2005 Meta-analysis: high-dosage vitamin E supplementation may increase all-cause mortality. Ann Intern Med 142:37–46

Morrow JD 2005 Quantification of isoprostanes as indices of oxidant stress and the risk of atherosclerosis in humans. Arterioscler Thromb Vasc Biol 25:279–286

Naasani I, Oh-Hashi F, Oh-Hara T et al 2003 Blocking telomerase by dietary polyphenols is a major mechanism for limiting the growth of human cancer cells in vitro and in vivo. Cancer Res 63:824–830

Neuhouser ML 2004 Dietary flavonoids and cancer risk: evidence from human population studies. Nutr Cancer 50:1–7

Neuhouser ML, Patterson RE, Thornquist MD, Omenn GS, King IB, Goodman GE 2003 Fruits and vegetables are associated with lower lung cancer risk only in the placebo arm of the beta-carotene and retinol efficacy trial (CARET). Cancer Epidemiol Biomarkers Prev 12:350–358

Pratico D, Tangirala RK, Rader DJ, Rokach J, FitzGerald GA 1998 Vitamin E suppresses isoprostane generation in vivo and reduces atherosclerosis in ApoE-deficient mice. Nat Med 4:1189–1192

Prieme H, Loft S, Nyyssonen K, Salonen JT, Poulsen HE 1997 No effect of supplementation with vitamin E, ascorbic acid, or coenzyme Q10 on oxidative DNA damage estimated by 8-oxo-7,8-dihydro-2′-deoxyguanosine excretion in smokers. Am J Clin Nutr 65:503–507

Rice-Evans C (ed) 2000 Wake up to flavonoids. Royal Society of Medicine Press, London

Richelle M, Turini ME, Guidoux R, Tavazzi I, Metairon S, Fay LB 1999 Urinary isoprostane excretion is not confounded by the lipid content of the diet. FEBS Lett 459:259–262

Sohal RS, Mockett RJ, Orr WC 2002 Mechanisms of aging: an appraisal of the oxidative stress hypothesis. Free Radic Biol Med 33:575–586

Spencer JP, Schroeter H, Rechner AR, Rice-Evans C 2001 Bioavailability of flavan-3-ols and procyanidins: gastrointestinal tract influences and their relevance to bioactive forms in vivo. Antioxid Redox Signal 3:1023–1039

Van Hoorn DE, Nijveldt RJ, Van Leeuwen PA et al 2002 Accurate prediction of xanthine oxidase inhibition based on the structure of flavonoids. Eur J Pharmacol 451:111–118

Velayutham M, Villamena FA, Navamal M, Fishbein JC, Zweier JL 2005 Glutathione-mediated formation of oxygen free radicals by the major metabolite of oltipraz. Chem Res Toxicol 18:970–975

Verhagen H, de Vries A, Nijhoff WA et al 1997 Effect of Brussels sprouts on oxidative DNA-damage in man. Cancer Lett 114:127–130

Virtamo J, Pietinen P, Huttunen JK et al 2003 Incidence of cancer and mortality following alpha-tocopherol and beta-carotene supplementation: a postintervention follow-up. JAMA 290:476–485

Wee LM, Long LH, Whiteman M, Halliwell B 2003 Factors affecting the ascorbate- and phenolic-dependent generation of hydrogen peroxide in Dulbecco's Modified Eagles Medium. Free Radic Res 37:1123–1130

Williamson G, Barron D, Shimoi K, Terao J 2005 In vitro biological properties of flavonoid conjugates found in vivo. Free Radic Res 39:457–469

Zhao K, Whiteman M, Spencer JP, Halliwell B 2001 DNA damage by nitrite and peroxynitrite: protection by dietary phenols. Methods Enzymol 335:296–307

DISCUSSION

Katan: Do you think flavonols from chocolate lower blood pressure in humans?

Halliwell: I'd have to go back and look at the design of the studies. The reason is that the act of eating lowers blood pressure, because blood goes to the gut.

Katan: There have been placebo-controlled studies, but they were mostly funded by Mars. It would be good to have some independent studies.

Aggett: I like a good bit of iconoclasm and we have had a good dose of that this morning. There is an important generic message from your presentation, and that is we should be using simple tests, assays and biomarkers that are well validated. The antioxidant field is plagued by a whole variety of tests that can be selectively chosen to get the required result. Can you comment on the overall quality of the assays and biomarkers of susceptibility to oxidative damage, and also the protective effect?

Halliwell: This question could support a symposium on its own! Let's take low density lipoprotein (LDL) oxidizability, as an example. You feed people vitamin E, isolate LDL, add some copper in the test-tube and you get a lag period that is lengthened with increasing dose of vitamin E. LDL oxidation in the test-tube doesn't really get going until the vitamin E is gone. Roland Stocker showed us nicely that in human atherosclerosis, lipid oxidation is going on and the vitamin E is still there. This tells me that these LDL oxidation studies are not representative of what is going on in the body. The free radical field has been plagued with people using ill-defined concepts such as oxidative damage, pro-oxidants and antioxidants. You have to really get down to the specific molecular level, both in terms of antioxidant action and in terms of what some of the reactive species do. Different molecular forms of oxidized lipids do very different things to cells. For example, if you take Alzheimer's disease brain and look for increased protein damage, this damage is focused on certain proteins. It is not random damage. Most of these damaged proteins are involved in energy metabolism, which fits nicely with the idea that in Alzheimer's neuronal energy metabolism is impaired. We have to get to this mechanistic mode instead of talking about total antioxidant capacity or vaguely about oxidative damage.

Stocker: With regard to general or systemic lipid oxidation, the best biomarker is now generally regarded to be F_2-isoprostane. If analysed properly, this is a useful maker. But 'analysed properly' refers to the need for a mass spectrometry-based method.

Aggett: How generalizable is F_2-isoprostane?

Stocker: There are two major issues. If you deal with a disease that is lipid-driven, such as atherosclerosis, it is important to standardize the F_2-isoprostane concentration to that of the lipid it is derived from, i.e., arachidonic acid, because the amount of that lipid may also change as a result of the disease. Some of the studies that have shown a benefit of vitamin E or other antioxidants on lipid oxidation *in vivo* have not done so in a scenario where lipid changes did occur. With regard to specificity, F2-isoprostanes are commonly measured in biological samples without distinction between different classes of lipids, and after the lipids have been hydrolysed. As a result, potentially important information is lost, such as the class of lipid (e.g. phospholipids or cholesteryl esters) the isoprostanes derived from.

Azzi: Barry Halliwell, I appreciate your courage in abandoning the traditional oxidants-antioxidant concepts. Similarly, one of the chief researchers in one of the major companies producing vitamins and carotenoids even proposed a symposium to be entitled 'Why the antioxidants have failed'. The antioxidant concept has been inflated and over-used, and there is very little evidence for this famous paradigm involving the bad guys (the radicals) and good guys (the antioxidants). I think it is appropriate to add a further comment regarding other molecules that have been considered antioxidants, such as flavonoids and polyphenols: their function as antioxidants is in most cases insignificant, due to their very low absorption and the lack of apparent recycling mechanisms to restore them after the modification produced by the radicals (Manach et al 2005). However, they are able to show other biological activities like modulation of gene expression or signal transduction (Rushmore & Kong 2002).

Halliwell: I agree with you. Some polyphenols are absorbed quite well, some quite badly. Many of them are COX2 inhibitors or can inhibit signal transduction. It's easy to see how this can happen in the gut. The big issue is whether people absorb enough into their body tissues for any effects to be significant, antioxidant or otherwise.

Manach: There is also a lot of interest in flavonoids as antioxidants. We must be suspicious of all *in vitro* studies in the field of flavonoids. These have almost always been done with high concentrations, and using compounds that are not present in the body. These compounds have no chance of getting to target tissues, and the doses used are much higher than the plasma concentrations achieved in the body (~1 µM). The tissue concentrations are even lower. I think that polyphenols probably don't have antioxidant action. Rather, we must look for gene transcription effects, signalling effects and induction of antioxidant defence systems by these compounds. To study this we must use small, physiologically relevant concentrations. We need to reinvestigate the *in vitro* studies on flavonoids, and we can't generalize to all polyphenols.

Boobis: If we look at the origins for some of the hypotheses of why these compounds might be effective, it is from observational studies on the effects of diet on health outcomes. The consumption of fruit of vegetables seems to be protective. But if one examines the evidence for effects of specific agents it becomes weaker. We have to keep in mind two entirely different possibilities. The first is that there is something present in our diet that is biologically active at the levels consumed, and it might be of advantage to identify this and develop it as a supplement. Such a compound would impact on physiological mechanisms. The evidence that we have identified specific agents with these properties is not very strong so far. There is a whole other dimension, though, which we shouldn't lose sight of. There are compounds in plants used as foods that we ingest but which have no effect at all on a normal individual. But if we take this same agent and give it at a pharmacological dose, it has a biological effect that is beneficial. It is quite different on the one hand to identify something from natural sources to develop as a supplement and on the other to produce something that can be given at a pharmacological dose. High levels might have an effect that is nothing to do with normal physiology but which may still be protective. The caveat here is this is essentially a drug and should be tested as such: demonstration of safety and efficacy is needed.

Halliwell: You raise an interesting point. When you go from the levels found in food to selling a 5 g supplement it may well be a drug.

Russell: Along those lines, in the ATBC study a pharmacological dose of β-carotene was used. Since then an animal model—a ferret exposed to smoke—has been studied. The ferret metabolises carotenes in much the same way as humans. High-dose β-carotene produced squamous metaplasia in smoke exposed ferrets. The metaplasia was much worse with the β-carotene plus smoke animals than in animals exposed to smoke alone. But with a physiological dose of β-carotene there was no effect at all.

Manach: In addition to the dose we should also consider the complexity of the food. It may be that a component doesn't have the same effect when it is isolated.

Azzi: We should ask ourselves a fundamental question. There are theories of disease and ageing based on radicals. If all these theories were valid why don't we find antioxidants capable of preventing all these diseases, including ageing?

Halliwell: Because most of the antioxidants used don't decrease free radical damage *in vivo*. This could be why they don't work!

References

Manach C, Williamson G, Morand C, Scalbert A, Remesy C 2005 Bioavailability and bioefficacy of polyphenols in humans. I. Review of 97 bioavailability studies. Am J Clin Nutr 81: 230S–242S

Rushmore TH, Kong AN 2002 Pharmacogenomics, regulation and signaling pathways of phase I and II drug metabolizing enzymes. Curr Drug Metab 3:481–490

Reduced folate status is common and increases disease risk. It can be corrected by daily ingestion of supplements or fortification

John M. Scott

School of Biochemistry and Immunology, Trinity College Dublin, Dublin 2, Ireland

Abstract. The natural folates are chemically unstable and poorly bioavailable in contrast to the chemical form, folic acid. Consequently most people, even those on good diets, have less than optimal nutrition with respect to this vitamin. Increased risks associated with deficiency include neural tube defects (NTDs) (proven), ischaemic heart disease and stroke (probable), certain cancers and decline in cognitive function (possible). Supplements of folic acid at the population reference intake (400 µg/d) completely normalizes all of these risks. Such levels are safe as judged by decades of use in wide sectors of the population. The main drawback of supplements is with respect to their effectiveness in preventing NTDs. They must be taken periconceptionally and before most women realise that they are pregnant. No more than one fifth of women take supplements effectively, largely due to the fact that over half of pregnancies are unplanned. This has led to the alternative of mandatory fortification of flour with folic acid in the USA and Canada. The level for such fortification is suboptimal for NTD prevention, because of fear of overexposure in the elderly. Thus even fortified communities require advice to take supplements for optimum NTD prevention.

2007 Dietary supplements and health. Wiley, Chichester (Novartis Foundation Symposium 282) p 105–122

The folates are the general name given to the eight reduced forms of the vitamin found in nature. The synthetic form folic acid used in supplements, fortified foods and for therapy is a provitamin and is not found in nature (Fig. 1). Folic acid is converted into the biologically active reduced form after its absorption and incorporation into cells. All but the circulating forms, mainly 5-methyltetrahydrofolate, exist in cells as a polyglutamate conjugates (Scott & Weir 1998). The reduced folate cofactors are essential in so-called carbon 1 transfers. Thus, tetrahydrofolate can accept the carbon 3 of the amino acid serine to give 5,10-methylenetetrahydrofolate (Fig. 2). This methylene group has three possible

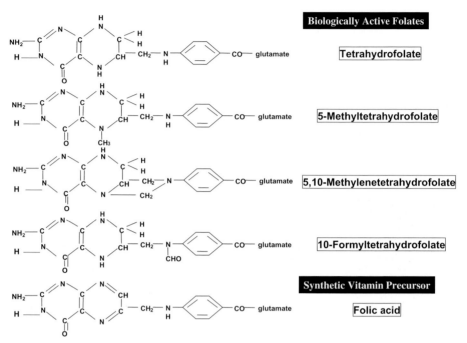

FIG. 1.   Structures of the most important naturally occurring folates and folic acid, the synthetic form of the vitamin.

fates. It can be used directly in pyrimidine biosynthesis by the key enzyme thymidylate synthase, to convert the uridylate-type base found in RNA into the thymidylate-type base found in DNA. It can be converted to 10-formyltetrahydrofolate, with this cofactor being used by two enzymes in the *de novo* purine biosynthetic pathway to enzymatically insert the carbon 2 and carbon 8 into the purine ring. Purine and pyrimidine biosynthesis both regenerate tetrahydrofolate, the former directly and the latter via the intermediate dihydrofolate (Fig. 2). As well as its role in *de novo* purine and pyrimidine biosynthesis 5,10-methylenetetrafolate can be reduced to 5-methyltetrahydrofolate. This form acts as a cofactor to the vitamin B12-dependent enzyme methionine synthase which, as can be seen from Fig. 2, is part of the methylation cycle. This cycle methylates homocysteine back to methionine. After its activation to S-adenosylmethionine the latter acts as the methyl donor to some three dozen methyltransferases that exist in all cells (Scott & Weir 1994). These enzymes have a wide range of functions, from methylating DNA, proteins and lipids, to the biosynthesis of creatine. One consequence of the methylation cycle is that all cells generate homocysteine from the S-adenosylhomocysteine. The latter is produced by the removal of the methyl group

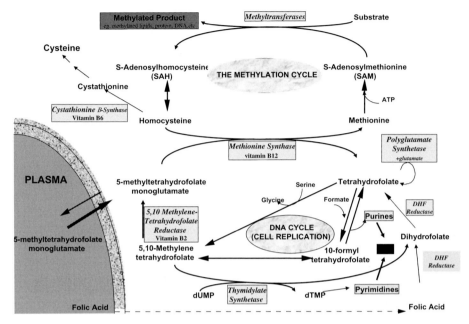

FIG. 2.   The role of the folates in the biosynthesis of purines and pyrimidines and thus DNA. Also shown is the methylation cycle and the key role played by the vitamin B12-dependent enzyme, methionine synthase.

of S-adenosylmethionine after its transfer to the methyl acceptors. In circumstances where there is reduced status of either folate or vitamin B12, both needed for methionine synthase, there is an inevitable accumulation of homocysteine. There is great variation in plasma homocysteine level between different individuals and populations depending upon their diet and the presence or absence of extra vitamins via fortification or supplementation. The four nutrients used by the three enzymes that utilize homocysteine are folate, vitamin B12 and vitamin B6, with vitamin B2 (Scott 2001). Increasing intake of vitamin B2 is important in those with a specific genetic variants, the $C \rightarrow T$ 677 polymorphism in the enzyme 5,10-methylenetetrahydrofolate reductase ($C \rightarrow T$ 677 MTHFR), (McNulty et al 2002, 2006). Elevated homocysteine due to reduced status for folate is very common but this can also happen if there is reduced status of vitamin B12 (Quinlivan et al 2002) or to a lesser extent vitamin B6 (Scott 2001).

There is a clear association between elevated homocysteine and risk of neural tube defects (NTDs), heart disease, stroke and certain cancers (Weir & Scott 1998). It is unclear if this risk is due to homocysteine *per se* or accumulation of its precursor S-adenosylhomocysteine, a potent inhibitor of all methyltransferases. Alternatively it may be because it is a marker of reduction in the status of folate or one of the other three vitamins. What is clear is that for all of the risk indications mentioned, including the complete optimization of total plasma homocysteine, a folic acid supplement at the physiological amount of folic acid 400 μg/d has a totally optimal effect as does a physiological amount of the other nutrients (Quinlivan et al 2002, McNulty et al 2002).

Concerning risk of excess intake, almost all supplements contain a maximum of 400 μg/d of folic acid. Historically, supplements of 5.0 mg were used but in modern times these would only be for short duration in the treatment of severe folate deficiency (Chanarin 1979). Even in these circumstances such levels as probably unnecessary with the ingestion of 400 μg/d being adequate to eliminate all deficiency, reduced status and risk. The higher levels of 5.0 mg/d are recommended to prevent the recurrence of NTD births and may in that limited circumstance be advisable (Department of Health 1992). Otherwise these high levels are unnecessary and unwarranted, producing no extra benefit and instead producing risk of preventing the timely diagnosis of vitamin B12 deficiency (Weir & Scott 1995).

Results and discussion

Folate deficiency

The historic emphasis with nutrients, particularly with respect to the micro-nutrients, was the prevention of deficiency. As mentioned above the folate cofactors are necessary for the *de novo* biosynthesis of purines and pyrimidines and thus DNA (Scott & Weir 1998). Thus it is not surprising that folate deficiency has been associated with reduced cell division. The most obvious clinical sign is a reduction in the cells of the bone marrow and seen as anaemia. The type of anaemia is specific, being characterized as a macrocytic, megaloblastic anaemia. This is because the red cells that are synthesized are larger than normal and this is seen as a raised mean corpuscular volume (MCV), in a full blood count (FBC). The megaloblastic nature of the anaemia refers to the appearance of abnormal red cell precursors called megaloblasts seen in bone marrow aspirates from folate deficient patients. An identical megaloblastic anaemia is also seen during vitamin B12 deficiency. This is known to arise as described in the now universally accepted methyl trap hypothesis (Scott & Weir 1994). *In vivo* the enzyme 5,10-methylenetetrahydrofolate reductase (MTHFR) is irreversible. Thus, once its product 5-methyltetrahydrofolate has been made in a cell the only way that it can subsequently be used is after its conversion back to tetrahydrofolate by the action of the vitamin B12-dependent enzyme methionine synthase. As the level of vitamin B12 decreases in cells so does the activity of methionine synthase, with an increasing inability to recycle back to tetrahydrofolate. Progressively this leads to more and more of the intracellular folates being metabolically trapped as the 5-methyl form. This leads in turn to a reduction not only in tetrahydrofolate but also in the carbon-one substituted folates, namely 10-formyltetrahydrofolate and 5,10-methylenetetrahydrofolate needed respectively for purine and pyrimidine biosynthesis. This gives rise to a form of 'pseudo' folate deficiency existing in vitamin B12-deficient cells. It thus explains in a very satisfactory way why the clinical presentation of vitamin B12 deficiency is mainly a megaloblastic anaemia that is identical to that seen in folate deficiency. Vitamin B12 deficiency can present with this anaemia, or with a very characteristic neuropathy. The latter is seen as a demyelination initially of the peripheral nerves but as it progresses also involving the main nerve tracts of the spinal column. The clinical presentation is a progressive ataxia leading eventually, if untreated, to paralysis and ultimately death. The full clinical condition is called sub-acute combined degeneration (SCD). At a biochemical level this neuropathy is almost certainly due not to the trapping of the folates as 5-methyltetrahydrofolate but more to the inability to supply methyl ($—CH_3$) groups to the methylation cycle. It seems clear that one of the methyltransferases is involved in maintaining the integrity of the myelin sheet. While vitamin B12-deficient cells contain an abundance of 5-methyltetrahydrofolates, these methyl groups cannot be fed into this methylation cycle (Fig. 2).

It is not clear why folate deficiency does not usually cause a similar neuropathy to vitamin B12 deficiency because its reduction would also be expected to reduce the methylation cycle's ability to function. One study in which folate deficiency was prolonged and severe did report neuropathy (Manzoor & Runcie 1976).

Folate deficiency severe enough to cause the anaemia described above is uncommon in most developed communities. It would be seen in those on very poor diets either through severe social deprivation but more frequently associated with chronic alcohol abuse. Historically, folate deficiency was also associated with the third trimester of pregnancy. This was conventionally thought to be due to a progressive transfer of maternal folate to the fetus leading to deficiency in the mother. It has now been shown that pregnancy accelerates in the rate of the body's catabolism (breakdown) of the vitamin (McPartlin & Scott 1997). Usually women on good diets, even in the absence of taking supplements or foods fortified with folic acid, have adequate stores to prevent them from becoming anaemic before the end of pregnancy.

Reduced folate status

General improvement in food intake in developed countries has seen to it that the prevalence of anaemia has become rare. This very much coincided with the view held by most nutritionists and dieticians that availability of a good mixed diet in most countries had eradicated deficiency. The assumption was, and with some still is, that a good mixed diet, most graphically depicted in the dietary pyramid, would not only have sufficient amounts of the macronutrients, (proteins, carbohydrates and fats) but would also supply all of the vitamins and trace elements.

It is now becoming clear that while this may well be true of some nutrients where dietary supply significantly exceeds requirements it is certainly not true of folate. In the case of folate, the main reason for this is that the multiple forms of the reduced derivatives of the vitamin found in the diet are chemically very unstable. In the usual steps of harvesting, distribution and storage very significant losses occur. Of even greater importance is that for many important sources of folate in the diet, methods of cooking and preparation leads to further significant losses (Hannon-Fletcher et al 2004). Finally, during their digestion and absorption from the diet there are further losses because nature's folates are not totally bio-available (Cuskelly et al 1996). This lack of bioavailability is probably due again to chemical instability rather than to the inability to convert the naturally occurring folate polyglutamate conjugates to their corresponding monoglutamate, a process necessary for their intestinal absorption. In any event, it is clear that efforts to increase status by improving dietary intakes are difficult. This is apparent from one of our own studies when we looked at the responses in normal healthy women (Cuskelly et al 1996). We compared four different results of intervention compared to no intervention (Fig. 3A). The measured outcome was the change in red cell folate level, as the best indicator of improvement in folate status (Fig. 3B). Dietary advice with a view to increase daily intake by 400 µg/d was totally ineffective. More importantly, an extra 400 µg/d of food folate showed no significant increase in folate status over the 20 weeks of the study, despite the fact that this extra food folate taken under daily supervision. Diets selected to contain an extra 400 µg of folic acid per day by way of fortified foods were successful. However, the most successful regime was the addition to the diet of one supplement of 400 µg/d. All four folate interventions had statistically improved intake (Fig. 3A). However, only fortification or supplements changed status (Fig. 3B). The reason for the success of the latter two regimes when compared to a similar supervised intake of 400 µg/d from food was that the former two regimes contained folic acid. This is the synthetic form of the vitamin (Fig. 1) that, in contrast to the natural folate, is extremely stable chemically and probably totally bioavailable (Scott & Weir 1998).

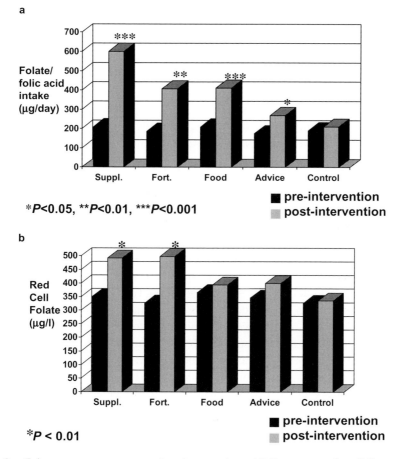

FIG. 3.   Folate status response to various interventions. (A) Response to four different types of dietary intervention on changes in folate/folic acid intakes compared to controls (µg/d). (B) The corresponding response in folate status as measured by changes in red cell folate levels (µg/litre) over the three week period of the intervention. Values shown as means in both instances. *$P < 0.05$, **$P < 0.01$, ***$P < 0.001$ (paired t-test). Data taken from Cuskelly et al (1996).

Benefits associated with supplements or fortified foods

Neural tube defects.   Between days 21 and 27 post conception, the neural plate closes to form the neural tube, which will eventually be the spinal column in the adult (Scott et al 1994). At the same time, a protrusion is sent out to cover the cranium. Improper closure of the neural tube leads to spina bifida with varying states of disability. Improper formation of the cranium results in anencephaly, which is incompatible with life. The two conditions combine about equally to account for nearly all of what are called neural tube defects (NTDs). An increased recurrence risk and the fact that different ethnic groups have very different prevalences for NTDs, shows that there is a genetic component. It is also clear that maternal nutrition is involved (Scott et al 1990).

A series of intervention trials have produced incontrovertible proof that the majority of NTDs can be prevented by the maternal periconceptional ingestion of folic acid (MRC Vitamin Study Research Group 1991). It was recommended that the increased intake to prevent occurrence of 400 µg/d could be achieved in any one of three ways: (i) a food folate, (ii) food fortified with folic acid, and (iii) supplements (Department of Health 1992). Largely as a result of our studies (Cuskelly et al 1996) showing the difficulty of changing status with food folate in practice, (Fig. 3) most campaigns changed with time to concentrate on getting women to take the extra folate/folic acid either in fortified foods or in supplements. However, it became clear over the decade and a half since then that the objective of taking folic acid correctly was only being achieved in about 20% of women (Botto et al 2005). This is largely due to the fact that in most communities over half of pregnancies are unplanned. Even in the planned pregnancies there is the problem of getting the information across and achieving compliance. This lack of success led the FDA to introduce fortification of flour with folic acid, this measure being mandatory in the USA and Canada by 1998. The most contentious issue was, and remains, the level of fortification used. The target dose to prevent occurrence of deficiency, 400 µg/d, was based on the amounts used in some of the trials and was really quite arbitrary. Further trials to determine if a lower level would be effective would be unethical. The dilemma was that if you added sufficient folic acid to ensure that women with low flour intake achieved the agreed target of 400 µg/d, the intake on those on high flour intake would mean that many people would receive 1000 µg/d (1.0 mg/d). It was agreed that such amounts would prevent the timely diagnosis of pernicious anaemia and other types of vitamin B12 deficiency. This happens because large amounts of folic acid if taken daily treat the anaemia in such subjects, which is the usual way that it is recognized. This so-called masking of the anaemia allows the neuropathy that also accompanies such vitamin B12 deficiency to progress undiagnosed to a stage where it may be irreversible. This masking is known to be very dose dependent. It is generally agreed that it never happens at the amount present in usual supplements of 400 µg/d. However, there is generally agreement that it definitely happens at levels of 5.0 mg/d and may happen at levels of 1.0 mg/d. For this reason, expert groups (Institute of Medicine 1998) have agreed that 1.0 mg/d should be the Tolerable Upper Intake Level (UL). Monitoring of flour intake in the US suggested that at a level of fortification of 140 µg/10 g of flour virtually nobody would exceed the UL. This amount, it was anticipated, would produce a mean increase in the US population of 100 µg/d of extra folic acid (Pfeiffer et al 2005). It turned out that to ensure compliance with the regulations most flour manufacturers have been adding amounts in excess, even up to double that target (Pfeiffer et al 2005). An important question was what level of NTD prevention would one anticipate from these levels of increase? Two studies by our own group contributed to this. The first study (Daly et al 1995) showed that in the index pregnancy of the women who went on to have an affected child less than 8% were folate deficient. However, there was 10-fold increase in risk of an affected birth between women at the lower end compared to the upper end of the normal physiological range (Fig. 4) In the second study we determined that a daily ingestion of as little as 200 μg/d would move all women to a level of optimal protection of NTDs (Daly et al 1997). These two studies were used to model the likely protective effect of fortification. Even at the elevated level due to overage of 200 μg/d as a mean increase it was accepted that only about half of NTDs would be prevented. Sufficient data have now accumulated to determine the effectiveness of fortification. While an earlier study in the USA reported a disappointing drop of less than 20%, this was almost certainly due to incomplete ascertainment (Honeim et al 2001). Three subsequent studies from Canada where ascertainment is very good, showed a drop of prevalence of about half (Ray et al 2002, Persad et al 2002), levels that would have been predicted from our earlier

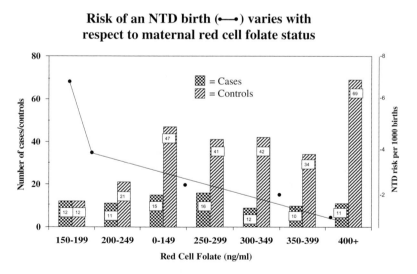

FIG. 4.   Risk of an NTD-affected birth defect from maternal folate status. The relationship between maternal folate status as determined by red cell folate (mean: μg/litre) in 84 NTD-affected pregnancies contrasted with 266 controls collected and analysed under the same conditions. The graph shows risk of an NTD pregnancy as a function of the comparative folate status. Data taken from the study by Daly et al (1995).

studies (Daly et al 1995, 1997). What was clear, however, was that because of the constraints of exceeding the UL, fortification alone could not be used as a reliable means to prevent all of the folate/folic acid preventable NTDs. A combination of other sources of voluntary fortification such as breakfast cereals or the daily inges-tion of supplements was needed in addition. Thus, while a fortification measure can prevent half of NTD affected births, for reasons of safety it must be combined with advice on other fortified foods or supplements.

Cardiovascular disease (CVD). There is a large volume of literature showing a very strong correlation between heart disease, venous thromboses and stroke, and eleva-tion of homocysteine in the circulation (Weir & Scott 1998). Inborn errors with marked elevations of homocysteine are associated with death from CVD. Recent meta-analyses have concluded on the basis of numerous retrospective but more recently prospective studies, that lowering plasma homocysteine by $3.0 \mu M/litre$ would cause an 11% reduction in ischaemic heart disease and a 19% reduction in stroke (The Homocysteine Studies Collaboration 2002, Wald et al 2002). Further analyses showed that having the abnormal polymorphism for the $C \rightarrow T$ 677 MTHFR, which increases plasma homocysteine, is associated with an increased risk of heart disease and stroke (Klerk et al 2002). These studies show associations. They have three interpretations (1) chance, (2) reverse causality (effect rather than cause) and (3) homocysteine is either a risk marker or a risk factor for these conditions. Which is correct can only be ascertained by intervention trials some of which have reported and some of which are ongoing (B-Vitamin Treatment Trialists Collaboration 2006). This same analysis showed that some earlier trials that found no association were greatly underpowered. This very important analysis of the power of past and future ongoing trials to detect an effect highlights the problem of nega-tive conclusions. It shows that the two large trials NORVIT and HOPE 2 that have been reporting negative findings are underpowered to detect the anticipated 11% reduction in heart disease (Bønaa et al 2006, The Heart Outcomes Prevention Evalu-ation [HOPE] 2 investigation 2006). Also, they have insufficient stroke events to reliably detect a 19% reduction. A further trial on stroke was also underpowered and took place after fortification with folic acid when homocysteine was greatly lowered in all involved (Toole et al 2006). The trial most likely to be definitive is the CHAOS-2 trial for CVD and the VITATOPS trial for stroke (B-Vitamin Treatment Trialists Collaboration 2006). These trials will not report until 2007 at the earliest.

Cognitive functions. There is an established literature showing a strong association between decreased cognitive function and high homocysteine levels (Clarke et al 2004). Also, homocysteine is thought to predict the rate of cognitive decline. Again, such homocysteine levels could be lowered by folic acid supplements or fortification.

Cancer. Folic acid is suggested to reduce the prevalence of some cancers (Ulrich & Potter 2006).

Risks associated with folic acid

Masking pernicious anaemia. As discussed above, this is really only agreed and established risk (Weir & Scott 1995). It is however felt that it will not occur at the level of ingestion of normal folic acid supplements of 400µg/d. The combination of supplements with fortified foods could move people above the UL and so they could become at risk (Scott & Browne 2006, Institute of Medicine 1998).

Cancer. One of the most effective anti-cancer drugs is the antifolate methotroxate. There is good theoretical grounds for concern and some animal evidence supporting the concept, that high levels of folic acid could not so much cause cancer but accelerate the growth of tumours (Ulrich & Potter 2006). To counterbalance this, one must consider that there is very good evidence that over one-third of the US population have taken multivitamins that contain in most instances 400µg/d of folic acid. This has been the practice for the last three or four decades (Ulrich & Potter 2006). No obvious connection between the taking of supplements and the emergence of cancers has emerged. Thus, while current concerns about this area cannot and should not be ignored, they can really only as of now be said to be theoretical. It is however certainly prudent to keep the levels of folic acid in supplements and in fortified foods either alone or combined to a minimum at the level that is thought to be effective. Folic acid is a provitamin and its conversion to the natural form is very dose dependent (Sweeney et al 2006). Exposure to high levels of unaltered folic acid would be a further cause for concern.

References

Bønaa KH, Njølstad F, Ueland PM et al 2006 Homocysteine lowering and cardiovascular disease events after acute myocardial infarction. N Engl J Med 354:1–9

Botto LD, Lisi A, Robert-Gnansia E et al 2005 International retrospective cohort study of neural tube defects in relation to folic acid recommendations-are the recommendations working? BMJ 330:571

B-Vitamin Treatment Trialists Collaboration 2006 Homocysteine-lowering trials for prevention of cardiovascular events: a review of the design and power of the large randomised trials. Am Heart J 151:282–287

Chararin I 1979 in the Megaloblastic Anaemias, 2nd edn. Blackwell Science, Oxford

Clarke R, Grimley-Evans J, Schneede J et al 2004 Vitamin B-12 and folate deficiency in later life. Age Ageing 33:34–41

Cuskelly CJ, McNulty H, Scott JM 1996 Effect of increasing dietary folate on red-cell folate: implications for prevention of neural tube defects. Lancet 347:657–659

Daly LE, Kirke PM, Molloy A, Weir DG, Scott JM 1995 Folate levels and neural tube defects. Implications for prevention. JAMA 274:1698–1702

Daly S, Molloy AM, Mills JL et al 1997 Minimum effective dose of folic acid for food fortification to prevent neural tube defects. Lancet 350:1666–1669

Department of Health 1992 Folic acid and the prevention of neural tube defects. Report from an expert advisory group. Department of Health, London

Hannon-Fletcher MP, Armstrong NC, Scott JM et al 2004 Determining bioavailability of food folates in a controlled intervention. Am J Clin Nutr 80:911–918

Homocysteine Studies Collaboration 2002 Homocysteine and risk of ischemic heart disease and stroke: a meta-analysis. JAMA 288:2015–2022

Honeim MA, Paulozzi LJ, Mathews TJ, Erickson JD, Wong LY 2001 Impact of folic acid fortification of the US food supply on the occurrence of neural tube defects. JAMA 285:2981–2986

Institute of Medicine 1998 Dietary reference intakes for thiamin, riboflavin, niacin, vitamin B_6, folate vitamin B_{12}, biotin, and choline. Washington (District of Columbia): National Academy Press

Klerk M, Verhoef P, Clarke R et al 2002 MTHFR 677 C $\rightarrow$ T Polymorphism and risk of coronary heart disease. JAMA 288:2023–3030

Manzoor M, Runcie J 1976 Folate-responsive neuropathy: report of 10 cases. BMJ 1:1176–1178

McNulty H, McKinley MC, Wilson B et al 2002 Impaired functioning of thermolabile methylenetetrahydrofolate reductase is dependent on riboflavin status: implications for riboflavin requirements. Am J Clin Nutr 76:436–441

McNulty H, Dowey LR, Strain JJ et al 2006 Riboflavin lowers homocysteine in individuals homozygous for the MTHFR 677 C $\rightarrow$ T polymorphism. Circulation 113:74–80

McPartlin J, Scott JM 1997 Identification and assay of folate catabolites in human urine. Methods Enzymol 281:70–77

MRC Vitamin Study Research Group 1991 Prevention of neural tube defects: results of the Medical Research Council vitamin study. Lancet 338:131–137

Persad VL, Van den Hof MC, Dube JM, Zimmer P 2002 Incidence of open neural tube defects in Nova Scotia after folic acid fortification. CMAJ 167:241–245

Pfeiffer CM, Caudill SP, Gunter EW, Osterloh J, Sampson EJ 2005 Biochemical indicators of B vitamin status in the US population after folic acid fortification: results from the National Health and Nutrition Examination Survey. Am J Clin Nutr 82:442–450

Quinlivan EP, McPartlin J, McNulty H et al 2002 Importance of both folic acid and vitamin B_{12} in reduction of risk of vascular disease. Lancet 359:227–228

Ray JG, Meier C, Vermeulen MJ et al 2002 Association of neural tube defects and folic acid food fortification in Canada. Lancet 360:2047–2048

Scott JM 2001 Modification of hyperhomocystinuria. In: Carmel R, Jacobsen DW (eds) Homocysteine in Health and Disease, Cambridge University Press, UK p 467–476

Scott JM, Weir DG 1994 Folate/Vitamin B_{12} interrelationships. Essays Biochem 28:63–72

Scott JM, Weir DG 1998 Folic acid, Homocysteine and one carbon metabolism: a review of the essential biochemistry. J Cardiovasc Risk 5:223–227

Scott JM, Browne P 2006 ANAEMIA/Megaloblastic Anaemia. In: Sadler MJ, Caballero B, Strain JJ (eds) Encyclopaedia of Human Nutrition, Second Edition, Elsevier, p 109–117

Scott JM, Kirke PM, Weir DG 1990 The role of nutrition in neural tube defects. Annu Rev Nutr 10:277–295

Scott JM, Weir DG, Kirke P 1994 Folate and neural tube defects. In: Bailey LB (ed) Folate in health and disease. Marcel Dekker, New York p 329–360

Sweeney MR, McPartlin J, Weir DG, Daly L, Scott JM 2006 Postprandial serum folic acid response to multiple doses of folic acid in fortified bread. Br J Nutr 95:145–151

The Heart Outcomes Prevention Evaluation (HOPE) 2006 Homocysteine lowering with folic acid and B vitamins in vascular disease. N Engl J Med 354–1-11

Toole JF, Malinow MR, Chambless LE et al 2006 Lowering homocysteine in patients with ischemic stroke to prevent recurrent stroke, myocardial infarction, and death: the Vitamin Intervention for Stroke Prevention (VISP) randomized controlled trial. JAMA 291:565–575

Ulrich CM, Potter JD 2006 Folate supplementation: too much of a good thing? Cancer Epidemiol Biomarkers Prev 15:189–193

Wald DS, Law M, Morris JK 2002 Homocysteine and cardiovascular disease: evidence on causality from a meta-analysis. BMJ 235:1–7

Weir DG, Scott JM 1995 The biochemical basis of the neuropathy in cobalamin deficiency. Baillieres Clin Haematol 8:479–497

Weir DH, Scott JM 1998 Homocysteine as a risk factor for cardiovascular and related disease: nutritional implications. Nutr Res Rev 11:311–338

DISCUSSION

Rayman: You didn't mention the effects of B6 and B12. Where do these come in?

Scott: The enzymes involved in controlling homocysteine metabolism have four cofactors: folate, vitamin B2, B6 and B12. If the level of any of these is compromised the level of homocysteine goes up. In the USA, now that the folate effect has been eliminated, the driver of homocysteine is vitamin B12. If you did an analysis before fortification in the USA, most of the variation in homocysteine was due to folate. If folate was 10, vitamin B12 would have been 2, and riboflavin would only be a player if it was TT, and B6 would be 1. B12 becomes a big player when you get rid of the folate effect. It still potentially could be a big player. It is of interest because of potential fortification and so on. We then could get into this whole area of why not add B12 to the diet and you wouldn't mask pernicious anaemia. The reason is that you wouldn't absorb B12 if you have pernicious anaemia.

Aggett: There is an intriguing possible effect with cancer of the colon. This is the dual risk/benefit analysis of folic acid in the risk of colon cancer in people who have early evidence of adenomatous polyp disease. In other words, if folic acid is given before there is any evidence of this, folic acid may be protective, but if there is evidence of this propensity then folic acid supplements are not protective.

Scott: Everyone figures out that the natural development of colon cancer is 20 or 30 years back, and can we look at this? I suppose the Nurses' Health Study (Giovannucci et al 1998) showed that supplement taking seemed to reduce colon cancer. This looked to be internally controlled. In the Nurses' Study it was only the nurses taking supplements for more than 15 years who showed real protection. If you are looking at the development of colon cancer, there is this issue: it is not amenable to investigation in a direct way. However, polyp recurrence is, so what you do then is look at someone who has polyps. They will need a repeat colonoscopy, so you put them into a clinical trial. This has been done with fibre. The studies looking at this are coming out with mixed results. One study doesn't show any reduction, but it doesn't show an increase. Those are important studies because they also address the issue of toxicity to some extent. It would be worrying if, in a controlled environment over a short period, we were to see increased polyps.

Aggett: The level of intervention is about 1000 µg, isn't it?

Scott: Yes. One of the issues that is a little worrying is that the conventional wisdom some five years ago was that folic acid is converted in cells into 5-methyltetrahydrofolate. In fact, it is converted when it passes the gut wall into this natural form. We started looking at this. It was thought to be dose dependent: if you take milligram amounts you get unaltered folic acid, but if you take small amounts you don't. It turns out that if you take smaller amounts you still get unaltered folic acid (Sweeney et al 2006). This is a concern. It is worrying that the levels of folic acid in the US food chain are so high. They don't need to be this high: beneficial effects probably occur with 200 µg. But because it is a food additive and not a supplement tablet it's hard to control the dose. It comes down to the question of caution: someone could turn round and say where is the evidence of any cancer risk? I was at the CDC meeting about 10 years ago. It was a contentious meeting, and some people in CDC wanted to double the level of fortification in the US diet. I spoke against it. A prominent person in CDC pointed at me and said, 'If you listen to John Scott there will be children in wheelchairs who don't need to be in wheelchairs'. I got up at question time and said, 'I agree with you, and I don't need you to tell me that. If we double the level of fortification you won't prevent 50% of NTDs, you'll probably prevent 80% of NTDs, but at what risk?' Then we get into this absence of evidence for a risk. This is a real dilemma. If folic acid were a drug the responsibility would be to produce evidence of no risk.

Wharton: A lot of the evidence you have shown illustrates the problem of sequential prevalence rates, before and after an event, rather than a proper controlled trial. The famous example of this, that is taught in epidemiology lessons in the UK, is that when the Samaritans service (a phone counselling service for people who are feeling suicidal) was introduced into East Anglia, the suicide rate fell dramatically. This was a before and after event. It just so happens that at the same time, coal gas disappeared, to be replaced by North Sea gas so it was much more difficult to commit suicide with a gas oven. The before and after evidence concerning effects of folic acid fortification doesn't impress me too much. Why don't the epidemiologists organize big community studies with proper controls? The US figures from before and after folic acid showed a 20% reduction (Honein et al 2001), but in the UK between 1980 and 1990, with no interventions, there was a 50% fall (Botting 1995).

Scott: That's a good point. Subsequent to the US data being released where ascertainment was poor, there was good ascertainment in Canada. The evidence in Canada from different provinces showed a consistent 40–50% reduction. There was an intermediate reduction between 1996 and 1998, which would be biologically consistent (Ray et al 2002). The reduction in stroke post-fortification is just an association (Yang et al 2006). It is consistent, and if we didn't find this we would be worried. One of the things they did in CDC was to ask their statisticians whether, from the history of stroke in the USA and Canada, they could find any other place where there was a sudden acceleration. They couldn't.

Taylor: Your presentation nicely illustrated why a regulator isn't asked to do a risk–benefit analysis when the issue is food fortification. Fortification is a blunt instrument, and you are looking at public health protection. At the same time you are doing the best you can for the population at risk. The data you showed perfectly illustrated the difficulties. The question is whether a regulatory agency could truly balance the B12 issues against the NTDs and decide who you put at risk.

Scott: In Ireland this has been an acute issue, because historically and legally, termination is not allowed. In the UK, 80–90% of fetuses with NTD are terminated. What we would have in Ireland is cumulatively groups of people with spina bifida and we have an obligation to do something. One thing that is a reasonable compromise is a dual policy of saying that women with reduced status are likely to be most at risk (Daly et al 1995) and are also likely to be the ones who won't take supplements. So we add an amount of folic acid that we think we can live with. It is difficult to justify doing no fortification.

Taylor: This can't be a drug model, and sometimes people have tried to squeeze fortification into a drug model which presents many problems.

Yetley: Your comments raised a couple of issues we covered earlier in the meeting: population monitoring and the need for interpretable, validated biomarkers. A commonly used approach for monitoring the safety and effectiveness of folic acid fortification and supplementation has been the percentage reduction of NTDs. This overlooks the fact that percentage reduction is dependent on the baseline level of NTDs in the population of interest. The goal is to effectively eliminate folate-responsive NTDs to the greatest degree possible. In a country with a relatively high rate of folate-responsive NTDs, they would need to achieve a much higher percentage reduction than a country with a low rate of folate-responsive NTDs in order to demonstrate similar success in terms of minimizing folate-responsive NTDs. However, changes in NTD rates may not be the best monitoring end point to use. Unfortunately, we don't know the threshold incidence rate for folate-responsive NTDs. Another problem is that complete ascertainment of early-terminated pregnancies is difficult to monitor in many countries; and incomplete ascertainment tends to underestimate the actual effectiveness of an intervention (e.g. fortification, supplementation) programme. In general, monitoring is most effective when biomarkers that are early indicators of changing risk are used. For example, it may be useful to look at the relationship between the incidence of NTDs and serum or red blood cell folate levels. Unfortunately, these relationships are not well documented. Moreover, there are virtually no data to aid in interpreting the use of serum or red blood cell folate levels for estimating risk of adverse effects other than NTDs. If, however, these relationships were well established, we could monitor changes in the serum and red blood cell folate levels and estimate the potential for changes in risk of various population groups. Then, if we have exceeded the folate status needed for benefit, food supply levels could be adjusted downward—and vice versa. However, until we have sufficient evidence that quantitatively links these biomarkers to functional end points, it is difficult to know how to use population monitoring data to decide on the safety and effectiveness of fortification and supplementation programmes.

Scott: Unfortunately, the only study that looked at risk was ours, because we had the historical samples (Daly et al 1995). The difficulty is to try to get correlation with protection where the outcome is NTDs now that there is a known protective effect. The MRC trial used 5 mg folic acid, which is an enormous amount, and still 20% of women got NTDs. There must be a proportion of NTDs that have nothing to do with folate, or which are intractable to it. This will be the baseline. If we look at the Canadian population, the evidence there is that there was 50% reduction. This was an impressive thing to do, and I think the FDA and CDC had the guts to do something, which was a courageous thing to do. It was a compromise and they knew there was an absence on good information on risk. But they still did it. The compromise that was reached in the USA and Canada, that they would prevent a proportion of preventable NTDs, was a step towards doing what they did.

Boobis: Can I explore the evidence for the link with cancer? You mentioned that one of the strands is the effects of anticancer drugs such as methotrexate. It seems to me that this doesn't provide secure evidence. They are not counterparts. You hit an enzyme which is critically involved in cell proliferation, those cells stop proliferating and die, and you stop the cancer developing. To argue the opposite, that adding in the cofactor means it will cause cancer, this is not well based in what we know about cell proliferation.

Scott: That is a valid point. It is a speculation. I don't believe that low-dose folic acid causes cancer. I don't think it would be proper for me to give a presentation like this, though, and to not mention this as a risk. People need to address this and hear about it in an informed group like this. If it is true, then it's a serious issue, even though there is just a small chance it is. One of the things that happens in cancer is that methotrexate is sometimes used followed by folic acid rescue. To this day people don't know how methotrexate works. When I was a young biochemist we were going to be the future, because we were going to design drugs that fitted into enzymes, but when they did the X-ray crystallography of dihydrofolate reductase, they found that methotraxate bound upside down. This illustrates how little we know about modes of action.

Halliwell: It is good to compare folic acid with vitamin E, where there is some suggestion of harm (which I find hard to believe), but little evidence of benefit.

Katan: I'd like to ask you about the bioavailability of folate from foods. In the USA, folate bioavailability is considered to be 50% of that of folic acid when taken alone. The basis for this is a paper by Sauberlich et al (1987), but we have looked at this and the data do not support this figure of 50%.

Scott: There are other studies. We did one where we looked at the recommendations to prevent NTDs (Cuskelly et al 1996). We had a group of normal women, which we split into a placebo group that got no extra folic acid and a group who were told to take folate rich foods. There was another group who took natural folates in foods, and we gave them 400 µg extra folate per day. There was a group that took 400 µg of folic acid and a further group that took 400 µg in fortified foods. The folate intake level increased in all of the latter three groups if you looked at dietary intake. Over the six week period there was no change in red cell folate in those getting the food, but there was in those getting the supplement and those eating fortified foods. This study wasn't designed to address bioavailability, but it does support a fairly big difference. A lot of bioavailability studies are not that well controlled, but it is probably not related to conjugation/deconjugation but rather stability (Hannon-Fletcher et al 2004). The people at the food company Birds Eye snap freeze their peas. They compared the folate content in peas that were snap frozen with those harvested and sold in a vegetable shop. There was a 90% reduction in the folate content in the non-frozen peas. This is supposed to be taken into account in food tables, but I don't think it is.

Katan: I agree that getting 400 µg of folic acid equivalent from food is too hard and that it should be done in the form of a supplement. But we have recently done an isotope dilution study for bioavailability, so I went through the literature and found that there are very few studies that address bioavailability in a proper way. We found 80% bioavailability for folate from foods compared with folic acid (Winkels et al 2007). There was another study which showed a similar number (Brouwer et al 1999). I don't think this value of 50% is correct.

Scott: The food issue had another undesirable effect. It created the image in the minds of GPs and pharmacists that if someone is on a good diet they don't need to worry about supplementation. One issue that is important is that many women who have had children with spina bifida blame themselves for not eating properly, but this isn't true.

Ernst: What food is supplemented in north America?

Yetley: Enriched cereal grains: for example, enriched flour, enriched rice, enriched cornmeal.

Ernst: In this natural experiment, would it be possible to create a natural control group of people who don't eat these foods?

Yetley: Fortified cereal grains are ubiquitous in the national food supply; it is difficult not to consume fortified foods or foods containing a fortified ingredient (e.g. enriched wheat flour).

References

Botting B 1995 The health of our children Decennial supplement. DS No 11 OPCS. London HMSO, London, p 48–58

Brouwer IA, van Dusseldorp M, West CE et al 1999 Dietary folate from vegetables and citrus fruit decreases plasma homocysteine concentrations in humans in a dietary controlled trial. J Nutr 129:1135–1139

Cuskelly GJ, McNulty H, Scott JM 1996 Effect of increasing dietary folate on red-cell folate: implications for prevention of neural tube defects. Lancet 347:657–659

Daly LE, Kirke PN, Molloy A, Weir DG, Scott JM 1995 Folate levels and neural tube defects. Implications for prevention. JAMA 274:1698–1702

Giovannucci E, Stampfer MJ, Colditz GA et al 1998 Multivitamin use, folate and colon cancer in women in the Nurses' Health Study. Ann Intern Med 129:517–524

Hannon-Fletcher MP, Armstrong NC, Scott JM et al 2004 Determining bioavailability of food folates in a controlled intervention study. Am J Clin Nutr 80:911–918

Honein MA, Paulozzi LJ, Mathews TJ, Erickson JD, Lee-Yang CW 2001 Impact of folic acid fortification of the US food supply on the occurrence of neural tube defects. JAMA 285:2981–2986

Ray JG, Meier C, Vermeulen MJ, Boss S, Wyatt PR, Cole DE 2002 Association of neural tube defects and folic acid food fortification in Canada. Lancet 360:2047–2048

Sauberlich HE, Kretsch MJ, Skala JH, Johnson HL, Taylor PC 1987 Folate requirement and metabolism in nonpregnant women. Am J Clin Nutr 46:1016–1028

Sweeney MR, McPartlin J, Weir DG, Daly L, Scott JM 2006 Postprandial serum folic acid response to multiple doses of folic acid in fortified bread. Br J Nutr 95:145–151

Winkels RM, Brouwer IA, Siebelink E, Katan MB, Verhoef P 2007 Bioavailability of food folates is 80% of that of folic acid. Am J Clin Nutr 85:465–473

Yang Q, Botto LD, Erickson JD, Berry RJ, Sambell C, Johansen H, Friedman JM 2006 Improvement in stroke mortality in Canada and the United States, 1990 to 2002. Circulation 113:1335–1343

Calcium and vitamin D

Kevin D. Cashman

Department of Food and Nutritional Sciences, and Department of Medicine, University College Cork, Cork, Ireland

Abstract. Calcium is required for normal growth and development as well as maintenance of the skeleton. Vitamin D is also essential for intestinal calcium absorption and plays a central role in maintaining calcium homeostasis and skeletal integrity. In addition, both micronutrients have important roles in non-skeletal-related physiological processes. Of concern, significant proportions of some population groups fail to achieve the recommended calcium intakes in a number of western countries. Furthermore, while cutaneous biosynthesis upon exposure of skin to ultraviolet B light is the major source of vitamin D for most people, this does not occur during winter time. Thus, there is an increased reliance on dietary sources during winter months to help maintain adequate vitamin D status. Since vitamin D is found naturally only in a limited number of foods, the usual dietary vitamin D intake by many European populations is not sufficient to maintain adequate vitamin D status. This paper will briefly review these important issues together with consideration of the potential role for supplementation with calcium and vitamin D in terms of improving their intakes in the population. It will also focus on the issues of efficacy as well as safety considerations of supplements.

2007 Dietary supplements and health. Wiley, Chichester (Novartis Foundation Symposium 282) p 123–142

Nutritional supplements constitute a sector of the dietary supplement market that is growing fast in all countries of the developed world. In this rapidly expanding market, vitamins and minerals constituted 51% of sales in 1997 (Nesheim 1999). There is a need however for unbiased, peer-reviewed evaluations that assess the scientific evidence as to the efficacy and safety of dietary supplements (Nesheim 1999). Efficacy can be defined as the ability of a supplement to provide a health benefit related to either the prevention of a nutritional deficiency or reduction in the risk of disease, while safety denotes a reasonable certainty that there will be no adverse effects from excessive intake of a nutrient (Hathcock 1997).

The present article will begin by briefly reviewing the physiological roles of calcium and vitamin D, their food sources and dietary requirements as well as the consequences and prevalence of their deficiency. It will then consider the potential role for supplementation with calcium and vitamin D in terms of improving the intakes of these micronutrients in the population. It will also focus on issues of efficacy as well as safety considerations of supplements.

Physiological roles of calcium and vitamin D in the body

Calcium

The adult human body contains about 1200 g of calcium, which amounts to about 1–2% of body weight. Of this, 99% is found in mineralized tissues, such as bones and teeth, where it is present primarily as calcium phosphate providing rigidity and structure (Cashman 2002). The remaining 1%, found in blood, extracellular fluid (ECF), muscle, and other tissues, plays a role in mediating vascular contraction and vasodilation, muscle contraction, nerve transmission and glandular secretion (Institute of Medicine 1997).

Calcium is required for normal growth and development of the skeleton (Cashman 2002). During skeletal growth and maturation, i.e. until the age of the early 20s in humans, calcium accumulates in the skeleton at an average rate of 150 mg per day. During maturity, the body—and therefore the skeleton—is more or less in calcium equilibrium. From the age of about 50 in men and from menopause in women, bone balance becomes negative and bone is lost from all skeletal sites. This bone loss is associated with a marked rise in fracture rates in both sexes, but particularly in women. Adequate calcium intake is critical to achieving optimal peak bone mass and modifies the rate of bone loss associated with ageing (National Institutes of Health 1994).

Vitamin D

Dietary-derived vitamin D as well as that synthesized dermally (see below) is processed and hydroxylated in the liver to 25-hydroxyvitamin D ($25[OH]D_3$) In the kidney, $25(OH)D_3$ undergoes a further hydroxylation at the first carbon to form 1,25-dihydroxyvitamin D ($1,25[OH]_2D_3$), which is the biologically most active form of vitamin D (Institute of Medicine 1997). The biological effects of $1,25(OH)_2D_3$ can be grouped into two categories, the 'classical actions', mainly affecting calcium homeostasis, and the 'non-classical actions', which include functions unrelated to calcium metabolism, such the regulation of cell differentiation and proliferation, cellular growth, and the regulation of hormone secretion, amongst others (Holick 2004).

Requirements and recommendations for calcium and vitamin D

Calcium

Given the high proportion of body calcium which is present in bone and the importance of bone as the major reservoir for calcium, development and maintenance of
bone is the major determinant of calcium needs. Thus, unlike other nutrients, the
requirement for calcium relates not to the maintenance of the metabolic function of
the nutrient but to the maintenance of an optimal reserve and the support of the
reserve's function (i.e. skeletal function) (Heaney 1997). Calcium requirements vary
throughout an individual's life, with greater needs during the periods of rapid growth
in childhood and adolescence, during pregnancy and lactation, and in later life.

There is considerable disagreement on human calcium requirements, and this is reflected in the wide variation in estimates of daily calcium requirements made by different expert authorities. For example, expert committees in the USA, UK and EU
have established very different recommendations for calcium intake (European Commission 1993, Institute of Medicine 1997, Department of Health 1998) (see Table 1).

TABLE 1   Recommended daily calcium (Ca) intakes in the UK, EU and USA

UK RNI (1998)[a]		*EU PRI (1993)*[b]		*US AI (1997)*[c]	
Age group (years)	*Ca intake (mg)*	*Age group (years)*	*Ca intake (mg)*	*Age group (years)*	*Ca intake (mg)*
0–1	525	0–1	400	0–0.5	210
1–3	350	1–3	400	0.5–1	270
4–6	450	4–6	450	1–3	500
7–10	550	7–10	550	4–8	800
11–14 M	1000	11–14 M	1000	9–13	1300
15–18 M	1000	15–17 M	1000	14–18	1300
11–14 F	800	11–14 F	800	19–30	1000
15–18 F	800	15–17 F	800	31–50	1000
19–50	700	19–50	700	51–70	1200
>50	700	>50	700	>70	1200
Pregnancy	no increment	Pregnancy	no increment	Pregnancy	
				≤18	1300
				19–50	1000
Lactation	+550	Lactation	+500	Lactation	
				≤18	1300
				19–50	1000

[a] Reference nutrient intake (RNI) (Department of Health 1998).
[b] Population reference intake (PRI) (European Commission 1993).
[c] Adequate intake (AI) (Institute of Medicine 1997).
Recommendations refer to both males and females unless stated otherwise. M = males; F = females.

Much of this divergence arises due to different interpretations of available human calcium balance data. The higher recommendations in the USA derive from defining calcium requirements based on maximal calcium retention estimated from human calcium balance studies, i.e. that which results in the maximum skeletal calcium reserve (Institute of Medicine 1997).

Vitamin D

Human skin contains the cholesterol precursor, 7-dehydrocholesterol, in relatively large amounts. On exposure to sunlight this compound absorbs ultraviolet B (UVB) light (wavelength 290–315 nm), which leads to the formation of the thermodynamically unstable previtamin D_3, precholecalciferol. Previtamin D_3 undergoes a slow non-enzymatic isomerization at skin temperature to form the more thermodynamically stable vitamin D_3 structure, cholecalciferol. If sun exposure is sufficient, very little if any vitamin D is required from the diet during summer. When sunlight is limited, dietary intakes of vitamin D can make a significant contribution to vitamin D status, depending on the food choice. For example, at latitudes >42°N, production of vitamin D_3 in winter is virtually zero, because the zenith angle of the sunlight increases in the autumn and winter and consequently, the amount of solar ultraviolet radiation that reaches the Earth's surface is substantially reduced (Holick 2004).

The Recommended Dietary Allowance (RDA) is the average daily dietary intake level that is sufficient to meet the nutrient requirements of nearly all (97–98%) healthy individuals in each life-stage and gender group. However, the establishment of an RDA for vitamin D is problematic since sunlight exposure and intake levels necessary to minimize hypersecretion of parathyroid hormone (PTH) are difficult to quantify (Institute of Medicine 1997). There is some disagreement on dietary vitamin D requirements, and this is reflected in the differing estimates of daily vitamin D requirements made by different expert authorities (see Table 2). The EU Population Reference Intake (PRI) range was established to allow for the variety of sun exposure between individuals. In the UK, there is no Reference Nutrient Intake (RNI) for vitamin D for people aged 4–64 years based on a lack of evidence that individuals in this age group rely on dietary intake for adequate vitamin D status (UK Department of Health 1998). However, the RNI for people aged 65 and over is 10 µg/day, which was considered to help safeguard against vitamin D deficiency and its adverse effects on bone health (UK Department of Health 1998) (Table 2). In contrast, the USA has established adequate intake (AI) values for these various age-groups (Table 2).

TABLE 2 Recommended daily vitamin D intakes in the UK, EU and USA

UK RNI (1998)[a]*		EU PRI (1993)[b]		US AI (1997)[c]	
Age group (years)	vitamin D intake (µg)	Age group (years)	vitamin D intake (µg)	Age group (years)	vitamin D intake (µg)
0–0.5	8.5	0–1	10–25	0–0.5	5
0.5–3	7	1–3	10	0.5–1	5
4–6	—	4–6	0–10	1–3	5
7–10	—	7–10	0–10	4–8	5
11–14	—	11–14	0–15	9–3	5
15–18	—	15–17	0–15	14–18	5
19–64	—	18–64	0–10	19–50	5
>65	10	>65	10	51–70	10
				>70	15
Pregnancy	10	Pregnancy	10	Pregnancy	5
Lactation	10	Lactation	10	Lactation	5

[a] Reference nutrient intake (RNI) (Department of Health 1998).
[b] Population reference intake (PRI) (European Commission 1993).
[c] Adequate intake (AI) (Institute of Medicine 1997).
*No RNI for those aged 4–64 years.

Food sources of calcium and vitamin D (including contribution from supplements)

Calcium

Milk and milk products are the most important dietary sources of calcium (contributing 51–73% of total calcium intake) for most people in western countries, with cereal products and fruits and vegetables each making a much smaller contribution (Institute of Medicine 1997, Cashman 2002). Tinned fish, such as sardines, are rich sources of calcium but do not make a significant contribution to intake for most people. Generally foods of plant origin are not very rich sources of calcium. However, due to the level of consumption, foods of plant origin do make a significant contribution to total calcium intake. For example, in the USA, cereals contribute about 25–27% of total calcium intake (Park et al 1997) while in the UK cereals contribute about 25% of total calcium intake with about 14% from bread due to calcium fortification of white flour (Department of Health 1998).

Increased availability of calcium-fortified foods and dietary supplements containing calcium salts is leading to a wider range of rich dietary sources of calcium. Contributions from nutritional supplements or medicines may be significant for some people. For example, a recent survey of a representative sample of Irish adults (aged 18–64 years) showed that of those subjects who took supplements, 66%, on average, took supplements containing vitamin D, while 37%, on average, took supplements containing calcium (Kiely et al 2001). These subjects consumed on average 188 mg and 4.5 µg of calcium and vitamin D, respectively, per day from supplements.

Vitamin D

Vitamin D as either of its molecular species, vitamin D_2 (derived from plant foods) or vitamin D_3 (derived from animal foods), is rather sparsely represented in the diet (Institute of Medicine 1997). Oily fish such as salmon, mackerel and sardines contain high amounts of vitamin D_3. Cod liver oil is also an excellent source of vitamin D_3. Some meats may contain $25(OH)D_3$. We have recently reported that meat and meat products (30.1%), fish and fish products (14.8%), and eggs and egg dishes (9.1%) were the main contributors to mean vitamin D intake in a representative sample of 18–64 year old Irish adults (Hill et al 2004).

Fortified foods can also be a major contributor to dietary vitamin D (Institute of Medicine, 1997). Use of vitamin D-containing supplements can also make a major contribution to mean daily intake of vitamin D. For example, in Ireland, 15% of adults aged 18–64 years took a supplement containing vitamin D (Hill et al 2004). In supplement users, the mean intake of vitamin D from supplements (4.6 µg/d) was higher than that from food sources (3.0 µg/d) (Kiely et al 2001). The percent contribution of supplements to vitamin D intake was about twice as high in Irish adult women (12%) as in men (6.8%) (Hill et al 2004). Supplemental vitamin D use may also be higher in the elderly (Hill et al 2004).

Consequences and prevalence of deficiency of calcium and vitamin D

Calcium

Because of the small metabolic pool of calcium (less than 0.1% in the ECF compartment) relative to the large skeletal reserve, for all practical purposes metabolic calcium deficiency probably never exists, at least not as a nutritional disorder. If, on the other hand, there is a chronic calcium deficiency resulting from a continual inadequate intake or poor intestinal absorption of calcium, circulating calcium concentration is maintained largely at the expense of skeletal mass, i.e. from an increased rate of bone resorption. This PTH-mediated increase in bone resorption is one of several important causes of reduced bone mass and osteoporosis (National Institutes of Health 1994, Institute of Medicine 1997). Chronic calcium deficiency resulting from inadequate intake or poor intestinal absorption may also have some role in the aetiologies of hypertension, including pre-eclampsia (McCarron 1997) and colon cancer (Department of Health Nutritional 1998).

Low calcium status as reflected in reduced bone mass appears to be common in western countries (European Commission 1998). In the absence of reliable indicators of nutritional adequacy for calcium, estimates of calcium deficiency are based largely on adequacy of dietary intake relative to recommendations. This approach is complicated by the lack of agreement between expert groups on recommended calcium intakes. In practice, estimates of the proportion of the population in different countries with inadequate calcium intake are based on recommended intakes for the individual countries. Using this approach, it has been reported that a significant proportion of some population groups fails to achieve the recommended calcium intakes in a number of western countries (Cashman 2002). In Ireland, 27% and 10% of the adult female and male population, respectively, who do not take supplements, have mean daily intakes of calcium below the estimated average requirements (<AR; an a measure of adequacy of intake of a nutrient) (Kiely et al 2001).

Vitamin D

It is well-established that prolonged and severe vitamin D deficiency leads to rickets in children and osteomalacia in adults (Institute of Medicine 1997). While the aetiology of osteoporosis is multifactorial, it is believed that secondary hyperparathyroidism as a result of mild vitamin D deficiency is a significant contributing factor (Lips 2001). In addition, in recent years more attention has focused on the emerging evidence that a circulating $25(OH)D_3$ level below 50 nmol/l may be contributing to the development of various chronic diseases that are frequent in western societies, including cardiovascular disease, hypertension, diabetes mellitus, some inflammatory and autoimmune diseases, as well as certain cancers (Zittermann 2003, Holick 2004).

Low vitamin D intakes have been reported in a number of studies in European adolescents, adults and elderly (for review, see Zittermann 2003, Holick 2004). Not surprisingly, low vitamin D status (as assessed by serum $25[OH]D_3$) has been reported to be common in these subjects, particularly during winter (for review, see Zittermann 2003, Holick 2004; and see Table 3).

The prevalence of vitamin D deficiency is dependent on the cut-off level of serum $25(OH)D_3$ applied, a much debated topic at present (Lips 2001, Zittermann 2003). For example, in Irish female adolescents, the prevalence may vary from about 30% to 85%, depending on whether you consider less than 25 nmol/l or less than 50 nmol/l as the appropriate cut-off value (Hill et al 2006a; and see Table 3). There is also a higher prevalence of vitamin D deficiency in Irish female adolescents than any other age-grouping, irrespective of the definition applied (Hill et al 2006a), highlighting vitamin D deficiency as a potential public health problem in the young.

TABLE 3 The percentages of subjects classified as vitamin D adequate or deficient in both seasons according to suggested 25(OH)D cut-off levels

| | August/September 2002 | | | | | February/March 2003 | | | | |
| | Females | | | | Males | Females | | | | Males |
Serum 25(OH)D cut-off (nmol/l)	11–13y	23–50y	51–69y	70–75y	20–64y	11–13y	23–50y	51–69y	70–75y	20–64y
						(%)				
<12.5	0	0	0	0	0	0	0	0	0	0
12.5–25	0	0	0	0	0	30	0	2	17	0
25–50	9	4	16	19	7	55	35	32	47	33
>50	91	96	84	81	93	15	65	66	36	67

[Serum 25(OH)D: >50 nmol/l, replete; 25–50 nmol/l, mild deficiency; 12.5–25 nmol/l, moderate deficiency; <12.5 nmol/l, severe deficiency]. Modified from Hill et al (2006a).

Nutritional and health benefits of calcium and vitamin D supplementation

The dietary deficiency of calcium and/or vitamin D identified in some population groups may be addressed in a number of ways. This includes changing eating behaviour at the population level by increasing the consumption of foods that are naturally rich in calcium (e.g. milk and milk products) or vitamin D (e.g. oily fish), fortification of foods consumed by target groups, or increasing calcium and vitamin D intakes from supplements. These may be seen as complementary rather than alternative strategies and each has advantages and disadvantages. For example, the use of supplements can be effective in increasing intake in individuals who consume them regularly but it has limited effectiveness at the population level due to the poor compliance with supplement use.

As mentioned previously, efficacy can be defined as the ability of a supplement to provide a health benefit related to either the prevention of a nutritional deficiency or reduction in the risk of disease (Nesheim 1999).

Calcium

In Ireland, the percentage of women with mean daily intakes of calcium < AR falls from 27% in non-supplement users to 16% in supplement users (Kiely et al 2001). The percentage of Irish men with mean daily intakes of calcium < AR was similar in users and non-users of supplements (14% and 10%, respectively). However, the mean intake of calcium from nutritional supplements was lower in men (98 mg) than that in women (240 mg) (Kiely et al 2001). Obviously, calcium is included in supplements in variable levels ranging from very low levels to levels approaching one gram or more.

In relation to reduction in the risk of disease, there is considerable evidence that increasing calcium intake above that usually consumed in the diet may have benefits for the development and maintenance of bone, and may reduce the risk of osteoporosis in later life. The findings of many of these controlled calcium intervention trials have been reviewed (Institute of Medicine 1997, Department of Health 1998, European Commission 1998). It should be noted that most of this evidence derives from studies that used calcium supplements; however, increasing calcium intake from foods naturally rich in calcium or calcium-enriched foods has similar effects. Supplementation of the usual diet with calcium has been shown to (i) increase the rate of accrual of bone mass in children and adolescents, (ii) reduce the rate of bone loss in postmenopausal women, and (iii) (particularly when combined with additional vitamin D) reduce the rate of bone fracture in elderly women and men. Two recent meta-analyses of studies of the effect of calcium on bone mineral density (BMD) and fracture in postmenopausal women have concluded that increasing calcium (via supplementation) has a positive effect on BMD, and a trend toward reduction of vertebral fractures (Shea et al 2002, 2004). Identification of the minimum effective dose is difficult to establish from such calcium intervention trials due to the fact that such studies generally only use one, or more rarely two test doses.

For some years low intakes of dietary calcium have been implicated in the aetiology of colon cancer and protection against colon cancer has been suggested as a potential benefit of higher calcium intakes (British Nutrition Foundation 1989). However, the US Food Nutrition Board, after reviewing the available data, have concluded that the evidence to support a link between dietary calcium intake and colon cancer risk is inconsistent (Institute of Medicine 1997).

There is some evidence that dietary calcium supplementation of the usual diet may help reduce established high blood pressure. In a review of 22 randomized intervention trials, calcium supplementation was found to reduce systolic blood pressure modestly in hypertensive adults, but had no effect in normotensive adults (Allender et al 1996), while diastolic blood pressure was not altered in either group. A meta-analysis of 14 randomized controlled trials of calcium supplementation during pregnancy found that supplements of 1500–2000 mg/d of calcium may result in a significant lowering of both systolic and diastolic blood pressure (Bucher et al 1996). However, the randomised controlled trial of Calcium for Pre-eclampsia Prevention in over 4500 women found no effect of calcium supplementation on hypertension, blood pressure, or pre-eclampsia (Levine et al 1997), perhaps because the intake of the control group (980 mg) was already relatively high (Institute of Medicine 1997).

There is evidence that there is probably a threshold of habitual calcium intake below which blood pressure is responsive to calcium intake and although this threshold has not been established it is very likely below the threshold necessary for maximal skeletal retention (Cashman & Flynn 1999).

Vitamin D

Vitamin D-containing supplement use certainly appears to be effective at lowering risk of vitamin D deficiency. The prevalence of low vitamin D status is much lower in supplement users than non-users (Table 4). Furthermore, our data on vitamin D status during winter and summer in users and non-users of vitamin D supplements suggest a benefit of vitamin D supplementation even during summer (Fig. 1). This raises the issue of optimizing vitamin D status throughout the year. We recently found that use of vitamin D-containing supplements (as well as high calcium intake) by Irish postmenopausal women was positively and significantly associated with vitamin D status in winter (Table 5) (Hill et al 2006b). In all of these studies, the vitamin D supplements were of relatively low dose, typically the median level was 6–7 µg/d. However, these doses appeared to be effective at lowering, albeit not eradicating, the prevalence of vitamin D deficiency.

In relation to reduction in the risk of disease, the potential benefits of vitamin D supplementation on reducing fracture incidence has been the subject of meta-analyses. A recent meta-analysis of randomized controlled trails of the effect of

TABLE 4 The percentages of women classified as vitamin D adequate/sufficient or deficient/insufficient according to a number of suggested 25(OH)D cut-off levels.

Serum 25(OH)D (nmol/l)	Lips (2001)				Vieth (1999)			Heaney & Weaver (2003)	
	<12.5	12.5–25	25–50	>50	<25	25–40	>40	<80	>80
February/March 2002									
All women (n = 59)	0	7	46	47	7	29	64	85	15
Vitamin D non-supplement users (n = 32)	0	13	47	40	13	38	49	94	6
Vitamin D supplement users (n = 27)	0	0	44	56	0	19	81	74	26
August/September 2002									
All women (n = 48)	0	0	17	83	0	4	96	63	37
Vitamin D non-supplement users (n = 36)	0	0	22	78	0	6	94	67	33
Vitamin D supplement users (n = 12)	0	0	0	100	0	0	100	50	50

Lips (2001) [Serum 25(OH)D: >50 nmol/l, replete; 25–50 nmol/l, mildly deficient; 12.5–25 nmol/l; moderately deficient; <12.5 nmol/l, severely deficient]. Vieth (1999) [Serum 25(OH)D: >40 nmol/l, adequate; 25–40 nmol/l, marginally deficient; <25 nmol/l; severely deficient]. Heaney & Weaver (2003) [Serum 25(OH)D: <80 nmol/l, insufficient; >80 nmol/l sufficient].
Modified from Hill et al (2005).

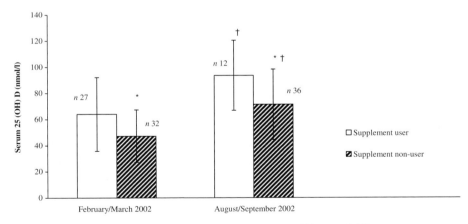

*Mean values significantly different from vitamin D-containing supplement users, within a season; $P \leq 0.01$.
†Mean values significantly difference from winter $P < 0.001$

FIG. 1. Serum 25(OH)D levels in 51–69 year-old Irish women (n = 59) stratified by vitamin D supplement use and by season. Modified from Hill et al (2005).

TABLE 5   Multiple linear regression analysis with serum 25(OH)D in 51–75 year-old women (*n* = 95) as the dependent variable

	B	95% CI	β	P value
Taking vitamin D supplements	12.918	2.943 to 22.892	0.255	0.012
Vitamin D intake	0.363	−0.333 to 1.059	0.990	0.303
Calcium intake	0.019	0.008 to 0.030	0.303	0.001
Sun habits, avoid sun	−1.871	−11.103 to 7.361	−0.036	0.688
Smoking, smokers	−22.303	−37.179 to −7.427	−0.272	0.004
BMI	−0.908	−1.936 to 0.120	−0.172	0.083
Age	−0.823	−1.489 to −0.157	−0.230	0.016

B, Coefficient; 95% CI, 95% confidence interval for B; β, standardized coefficient.

vitamin D supplementation on fracture prevention reported that vitamin D supplementation between 700 to 800 IU/d (17.5–20 µg/d) appeared to reduce risk of hip and non-vertebral fractures in elderly persons (Bischoff-Ferrari et al 2005). However, an oral dose of 400 IU/d (10 µg/d) did not appear to be sufficient for fracture prevention. In our study of a representative sample of Irish adults only 2.4% of subjects had an intake of 20 µg/d (Hill et al 2004).

Daily vitamin D supplementation (typically with 10 µg/d) is used as a prophylactic measure for rickets in many countries. While there is far less data available, there is some evidence for a beneficial effect of vitamin D supplementation on muscle function, rheumatoid arthritis, blood pressure, blood glucose and insulin levels (reviewed in Zittermann 2003). However, in many cases the doses were high (typically with 20–2500 µg/d) and in some cases in combination with calcium; low dose supplementation did not appear to be effective.

Absorption and bioavailability of supplemental calcium and vitamin D

Calcium

The bioavailability of calcium from various forms of supplemental calcium appears to be good. Sheikh et al (1987) reported that calcium absorption from a 500 mg calcium load as the acetate, lactate, gluconate, citrate or carbonate salts was similar (27–39%) in human subjects to that from whole milk (31%). In another comparative study, Recker et al (1988) found that fractional calcium absorption from a 250 mg calcium load in a group of 10 healthy postmenopausal women was as good from calcium carbonate (22%) as from dairy products, such as whole milk (27%), chocolate milk (23%), yoghurt (25%), imitation milk (22%) or cheese (23%).

There is some evidence of enhanced absorbability of calcium from highly soluble salts. Studies on calcium citrate malate (CCM), a highly soluble mixed salt (aqueous solubility ~80 mM) developed for food fortification purposes, suggest enhanced bioavailability of calcium from this source. Smith et al (1987) reported that calcium absorption from CCM taken as a tablet with a breakfast meal was higher (37%) than from a calcium carbonate supplement (29%) taken under similar conditions. Likewise, calcium absorption from CCM mixed with orange juice was higher than calcium absorption from milk (38% versus 29%). The bioavailability of vitamin D from supplements is known to be good (Institute of Medicine 1997).

Toxicity

Calcium

The available data on the adverse effects of calcium in humans are primarily from the intake of calcium from nutritional supplements. The three most widely studied and biologically important are: (i) kidney stone formation (nephrolithiasis); (ii) syndrome of hypercalcaemia and renal insufficiency, with or without alkalosis (referred to historically as milk alkali syndrome associated with peptic ulcer treatments); and (iii) effect on absorption of other essential minerals. These have been reviewed elsewhere (Institute of Medicine 1997, Flynn & Cashman 1999).

Based largely on the data on the association of high calcium intakes with hypercalcaemia and renal insufficiency in adults the US Food and Nutrition Board have established a Tolerable Upper Intake Level (UL) of calcium of 2500 mg/d for children, adolescents and adults, as well as pregnant and lactating women (Institute of Medicine 1997). UL is defined as the maximum level of daily calcium intake that is unlikely to pose a risk of adverse health effects in almost all individuals in the specified life-stage and gender group. The term connotes a level of intake that can, with high probability, be tolerated biologically (Institute of Medicine 1997). In Irish adults, only 0.3% of subjects had intakes of calcium above the UL.

Vitamin D

There are no reports in the literature about vitamin D intoxication with traditionally consumed foods. All reports of vitamin D intoxication involve either highly-fortified foods or pharmacological doses of vitamin D supplements (for review, see Vieth 1999). In almost every instance of vitamin D intoxication, serum $25(OH)D_3$ levels are >200 nmol/l (Vieth 1999, Zittermann 2003), but are usually well above this level (Vieth 1999). Vitamin D intoxication is characterized by hypercalcaemia and hypercalciuria, and can lead to the calcification of soft tissues, with minimal or no change in $1,25(OH)_2D$ levels (Vieth 1999).

The current No Observed Adverse Effect Level (NOAEL) for vitamin D is set at 100 µg/day (Scientific Committee on Food 2002). This figure is based on the consensus that at a vitamin D intake of 100 µg/day, the risk of hypercalcaemia/hypercalciuria starts to increase. The UL for vitamin D stands at 50 g/day

(Institute of Medicine 1997, Scientific Committee on Food 2002), which incorporates a safety factor of 2 from the NOAEL. In Irish adults, the 95th percentile of intakes of vitamin D from all sources (i.e. from foods and supplements) were well below the UL (Hill et al 2004).

Conclusion

Suboptimal intakes/status of calcium and vitamin D are common in many individuals. Nutritional supplements can certainly reduce the incidence of these nutritional deficiencies, but some consideration is needed in relation to the minimum effective dose of these micronutrients. Calcium and vitamin D may also lower the risk of certain diseases, but again the issue of minimum effective dose is relevant. In general, it would appear that the levels of supplementation which bring about reductions in disease risk are higher than those presently included in many supplements. Further research is needed to evaluate the effective doses while ensuring safety. In general, the level of inclusion of calcium and vitamin D in supplements at present appears to be sufficient so as not to place individuals at risk of excessive intake.

Acknowledgement

The preparation of this manuscript was made possible in part by funding received by the author from the Higher Education Authority, Dublin.

References

Allender PS, Cutler JA, Follmann D, Cappuccio FP, Pryer J, Elliott P 1996 Dietary calcium and blood pressure: a meta-analysis of randomized clinical trials. Ann Intern Med 124:825–831

Bischoff-Ferrari HA, Willett WC, Wong JB, Giovannucci E, Dietrich T, Dawson-Hughes B 2005 Fracture prevention with vitamin D supplementation: a meta-analysis of randomized controlled trials. JAMA 293:2257–2264

British Nutrition Foundation 1989 Calcium: the report of the British Nutrition Foundation's task force on calcium. British Nutrition Foundation, London

Bucher HC, Guyatt GH, Cook RJ et al 1996 Effect of calcium supplementation on pregnancy induced hypertension and pre-eclampsia: a meta-analysis of randomized controlled trials. J Am Med Assoc 275:1113–1117

Cashman KD 2002 Calcium intake, calcium bioavailability and bone health. Br J Nutr 87 Suppl 2:S169–177

Cashman KD, Flynn A 1999 Optimal nutrition: calcium, magnesium and phosphorus. Proc Nutr Soc 58:477–487

Department of Health 1998 Calcium. In: Dietary reference values for food energy and nutrients for the United Kingdom. H.M. Stationery Office London p 150–157

Department of Health Nutritional 1998 Aspects of the development of cancer. Report of the working group on diet and cancer of the committee on medical aspects of food and nutrition policy. H.M. Stationary Office London

European Commission 1993 Calcium. In: Nutrient and energy intakes of the European community. Report of the scientific committee for food (31st series), Luxembourg p 136–145

European Commission 1998 Report on osteoporosis in the European Community: action for prevention. Office for Official Publications for the European Commission, Luxembourg

Flynn A, Cashman K 1999 Calcium fortification of foods. In: Hurrel R (ed) Mineral Fortification of Foods. Surrey: Leatherhead Food RA p 18–53

Hathcock JN 1997 Vitamins and minerals: efficacy and safety. Am J Clin Nutr 66:427–437

Heaney RP 1997 The roles of calcium and vitamin D in skeletal health: an evolutionary perspective. Food, Nutrition and Agriculture No. 20:4–12

Heaney RP, Weaver CM 2003 Calcium and vitamin D. Endocrinol Metab Clin North Am 32:181–194

Hill TR, O'Brien MM, Cashman KD, Flynn A, Kiely M 2004 Vitamin D intakes in 18-64-y-old Irish adults. Eur J Clin Nutr 58:1509–1517

Hill T, Collins A, O'Brien M, Kiely M, Flynn A, Cashman KD 2005 Vitamin D intake and status in Irish postmenopausal women. Eur J Clin Nutr 59:404–410

Hill TR, Flynn A, Kiely M, Cashman KD 2006a Prevalence of suboptimal vitamin D status in young, adult and elderly Irish subjects. Ir Med J 99:48–49

Hill TR, O'Brien, Jacobsen J et al 2006b Vitamin D status of 50-75 year-old Irish women: its determinants and impact on biochemical markers of bone turnover. Public Health Nutr 9:225–233

Holick MF 2004 Vitamin D: importance in the prevention of cancers, type 1 diabetes, heart disease and osteoporosis. Am J Clin Nutr 79:362–371

Institute of Medicine 1997 Calcium. In: Dietary reference intakes: calcium, magnesium, phosphorus, vitamin D, and fluoride. National Academy Press, Washington DC 4/1–4/57

Kiely M, Flynn A, Harrington KE et al 2001 The efficacy and safety of nutritional supplement use in a representative sample of adults in the North/South Ireland Food Consumption Survey. Public Health Nutr 4:1089–1097

Levine RJ, Hauth, JC, Raymond EG, Bild DE, Clemens JD, Cutler JA 1997 Trial of calcium to prevent preeclampsia. New Engl J Med 337:69–76

Lips P 2001 Vitamin D deficiency and secondary hyperparathyroidism in the elderly: consequences for bone loss and fractures and therapeutic implications. Endocrinol Rev 22:477–501

McCarron DA 1997 Role of adequate dietary calcium intake in the prevention and management of salt-sensitive hypertension. Am J Clin Nutr 65:712S–716

National Institutes of Health 1994 Optimal calcium intake. NIH Consensus statement vol 12: no. 4. NIH, Bethesda, MD

Nesheim MC 1999 What is the research base for the use of dietary supplements? Public Health Nutr 2:35–38

Park YK, Yetley EA Calvo MS 1997 Calcium intake levels in the United States: issues and considerations. Food, Nutrition and Agriculture No. 20:34–43

Recker RR, Bammi A, Barger-Lux MS, Heaney RP 1988 Calcium absorbability from milk products, an imitation milk, and calcium carbonate. Am J Clin Nutr 47:93–95

Scientific Committee on Food 2002 Opinion on the tolerable upper intake level of Vitamin D. December 2002. SCF/CS/NUT/UPPLEV/38 Final

Shea B, Wells G, Cranney A et al 2002 Meta-analyses of therapies for postmenopausal osteoporosis. VII. Meta-analysis of calcium supplementation for the prevention of postmenopausal osteoporosis. Endocrine Rev 23:552–559

Shea B, Wells G, Cranney A et al 2004 Calcium supplementation on bone loss in postmenopausal women. Cochrane Database Syst Rev 1: CD004526

Sheikh MS, Santa Ana CA, Nicar MJ, Schiller LR, Fordtran JS 1987 Gastrointestinal absorption of calcium from milk and calcium salts. New Engl J Med 317:532–536

Smith KT, Heaney RP, Flora L, Hinders SM 1987 Calcium absorption from a new calcium delivery system (CCM). Calcified Tissue Int 41:351–352

Vieth R 1999 Vitamin D supplementation, 25-hydroxyvitamin D concentrations, and safety. Am J Clin Nutr 69:842–856

Zittermann A 2003 Vitamin D in preventive medicine: are we ignoring the evidence? Br J Nutr 89:552–572

DISCUSSION

Katan: Walter Willett has voiced concerns that very high calcium intake may be associated with prostate cancer (Giovannucci et al 1998). I saw some data from an independent set of epidemiological studies which also found this association (Gao et al 2005). Can you comment on this?

Cashman: I can't say much more than there are these two sets of data. Purely from a mechanistic perspective, $1,25(OH)_2D_3$ has a role in terms of prostate cancer. If you are increasing calcium intake, perhaps you are suppressing this through the PTH axis.

Katan: But is it real or causal?

Cashman: We've had a day and a half discussing the merits of epidemiological studies versus intervention. Purely from an ethical perspective, I don't think we are going to do an intervention study to look at increasing calcium intake to observe an increase in cancer risk. This needs to examined carefully in future calcium supplementation studies in men.

Russell: You mentioned supplementation of vitamin D being needed to get blood levels up to a certain point. Is there any thinking in Europe of fortification with vitamin D?

Cashman: There was a three year project looking at vitamin D fortification, funded by the EC. It became obvious quite soon that different European countries had contrasting views on supplementation practices, let alone on the issue of fortification. While some countries would be very receptive, there were others who wouldn't consider it warranted. The Finnish have been very innovative because they have looked at how we might fortify non-fat foods, rather than fortifying milk. They have developed a technology that permits the inclusion of vitamin D in a homogeneous way into something like bread. The issue of fortification with vitamin D at a European level is something which the European Commission might need to consider.

Alexander: In Norway we have had discussions about fortification. Presently we fortify milk and margarine, but we are discussing fortifying more foods. What I wanted to elaborate a bit more on was the difference between D2 and D3. I think we should pay more attention to this; so far I don't think supplements differentiate between these two types. D2 is in fact a pharmaceutical product different from the natural hormone, vitamin D3, synthetized in the skin and one needs much more D2 to have an effect (Houghton & Vieth 2006). You can even suppress the conversion of D3 to active metabolites by giving D2.

Cashman: In Ireland and the UK, many of the supplements have moved to include D3 as opposed to D2. In the USA, I thought that D2 was used for fortification because it is cheaper. The existing scientific evidence seems to suggest that D3 is more effective.

Manach: You have a strategy to increase calcium to prevent osteoporosis, but is there a strategy to eliminate the calcium elimination from bones?

Cashman: Yes, but in the relative scheme of things, it is much more likely that increasing calcium intake will have a much bigger impact on bone health than going about trying to conserve calcium. If you want to maintain calcium effectively you need to limit your salt intake and your caffeine intake. But increasing calcium intake will be a much more effective strategy for optimizing the availability of calcium to bone. It is the same issue that comes up in terms of bioavailability of supplements. The supplement industry has spent millions trying to find supplements that are more bioavailable, and for the most part the bioavailability is only modestly better. Instead of 30% they might be of the order of 35%, but this improvement in supply of calcium to the body is easily achieved by putting in a little extra of the mineral in the supplement/food.

Scott: After the Second World War vitamin D fortification was used in the UK, and there were some deleterious effects with some children. Was this because of high levels?

Wharton: The exact intakes in individual children were not recorded but calculations suggested that some children were receiving 100 µg (4000 IU) of vitamin D daily. Vitamin D was added to infant formula and to weaning cereals and many mothers were giving cod liver oil. This caused an epidemic of hypercalcaemia, which resulted in some cases in vomiting and failure to thrive, but the most severe forms resulted in brain damage. Here were mothers who thought they were doing well for their children by following the advice of healthcare professionals, and ended up poisoning their children (see review by Wharton & Darke 1982, British Paediatric Association 1956). We shouldn't be gung ho about supplements and fortification. It is a difficult balance.

I have a question. There is a fair amount written on deciding what a normal plasma $25(OH)D_3$ level is, such as whether it suppresses PTH below a certain level (Vieth & Walfish 2000). What do you think about this?

Cashman: In the adult population, to suppress PTH to flatline you need about 78 nM/litre of $25(OH)D_3$, which is high and quite difficult to achieve, especially during winter. To get to this level you need to really be taking vitamin D supplements in the winter. Sun exposure in the summer can get you to that level. About 80 nM/litre has been shown to stimulate calcium absorption in elderly people. The school of thought that 50 nM/litre may be a good cut-off for vitamin

D status is partly because it is very hard to get to 80 nM/litre, but also because levels of vitamin D below 50 nM/litre have been linked to increased risk of other chronic diseases as I discussed in my presentation. In the UK it is 25 nM/litre but this was always established in terms of prevention of rickets rather than osteoporosis.

Aggett: It is important that we have a physiological basis for setting the reference value for $25(OH)D_3$. I discovered recently that some chemical pathology laboratories operate with two reference ranges in the UK, one for winter and one for summer. Clearly, they thought there was some sort of logic to this. It comes back to the old idea of the misinterpretation of creating reference ranges and not putting them into appropriate physiological contexts. This is so common in nutrition. I also wanted to raise the point that we are not totally cavalier in the UK about not having a recommended intake of vitamin D. We encourage people to get some sunshine. But your comments mean that we should start thinking about the soundness of this advice. This is tempered by the fact that we still have vitamin D fortification for some of the traditional fat spreads, but not necessarily for the fat spreads that are replacing them. The other issue that we've had recently about vitamin D supplements for pregnant women and children under the age of five is that in the UK a public health arm has been amalgamated with NICE (National Institute of Clinical Excellence), which reasonably is very much oriented on evidence-based practice and is informed strongly by experience and practice in academic and secondary care. NICE recently commented that there is no good evidence for the use of these supplements, thereby compromising the public health message that has gone out about the routine use of these in pregnancy. This has generated a lot of active discussion between several agencies.

Cashman: In Ireland we are in a similar situation with prescribing vitamin D supplementation during pregnancy. Getting the evidence base is difficult because ethically you are not allowed to take the blood from healthy (non-ill) children. The UK recommendation is zero because there is a large dependence on sunshine exposure. However, we are being strongly advised by the skin cancer agencies to avoid sun exposure. Suncream with SPF 8 will largely eliminate vitamin D synthesis. This is where I have a concern: there needs to be a meeting of minds to put out some kind of a public health message that doesn't lead to problems with skin cancer but which gets individuals, especially children, to have some appropriate exposure to sunshine. There was a recent article in the British Medical Journal (Diffey 2005) that made simple recommendations such as encouraging children to walk to school on the sunny side of the road. With regard to fortification of spreads/margarine, in Ireland it moved from mandatory fortification to voluntary fortification, so we don't know which the manufacturers are including vitamin D.

Rayman: Moving to the elderly, there are good data from the USA showing that vitamin D levels are related to the progression of osteoarthritis. There are a number of recent papers from Michael Holick looking at risk of cancer (Holick 2006). One of the things he talked about is that there is a 17% reduction in cancer incidence and a 29% reduction in cancer mortality for an increment of 25 nM/litre of 25(OH)D$_3$. The evidence seems to be growing on the cancer front.

Cashman: Larger increments in serum 25(OH)D$_3$ concentrations, which appear to be related to reduction of risk of some diseases, are difficult to achieve by dietary means. So do we ask people to take supplements, or do we explore the fortification mode?

Rayman: If we look at 25(OH)D3 levels in Europe the only country that is coming out well is Norway, and that presumably is because of fish eating. We all fall quite low in winter.

Russell: There is an older study from Denmark on veiled Muslim women taking in about 15 μg of vitamin D per day and still having very low levels of circulating 25(OH)D. This implies that in this population, with no exposure to sunlight, that the need for ingested vitamin D is quite large. Are any public health considerations being given to European Muslim populations?

Cashman: One of the cohorts in our intervention trial was Muslims living in Copenhagen. They were strict Muslims. Many of these individuals have extremely low levels of vitamin D, which are typical of people with rickets. There is some evidence that they are getting very low exposure to sun during summertime, as a consequence of their traditional garb, and because they have darker skins they are more resistant to synthesizing it. You need quite high supplement levels even to get them out of the risk of having vitamin D levels below 30 nM/litre.

Przyrembel: Does anyone have quantitative data on what skin surface should be exposed to sun, for how long, to produce how much vitamin D?

Rayman: 15 min for the forearms and face have been proposed, but it depends where the sun is and what latitude you are at.

Katan: In the Netherlands almost half the African and Asian immigrants are vitamin D deficient, and quite a number of them are really severely deficient. They don't use margarine, so I thought what would be needed to be fortified would be the olive oil.

Rayman: Do they drink milk?

Katan: Very little. They eat some yoghurt. Anyway, the dairy industry isn't about to add something to milk in the Netherlands.

Russell: Calcium supplements were mentioned, and that it doesn't matter which one is used. Isn't there evidence that in elderly people, calcium carbonate is relatively less bioavailable than calcium citrate?

Cashman: In a situation where there is low gastric acid, calcium carbonate is less absorbed because it isn't very soluble. This is in a particular situation where elderly subjects have achlorhydria. If you are talking about more general use of the adult population there is very little difference between them. There is some evidence that calcium citrate malate is better, but it is only a small difference, and it is possibly as cost effective to add a little more calcium carbonate.

References

British Paediatirc Association 1956 Hypercalcaemia in infants and vitamin D. Br Med J 2:149–151

Diffey B 2005 Do white British children and adolescents get enough sunlight? Br Med J 331:3–4

Gao X, LaValley MP, Tucker KL 2005 Prospective studies of dairy product and calcium intakes and prostate cancer risk: a meta-analysis. J Natl Cancer Inst 97:1768–1777

Giovannucci E, Rimm EB, Wolk A et al 1998 Calcium and fructose intake in relation to risk of prostate cancer. Cancer Res 58:442–447

Holick MF 2006 Vitamin D: its role in cancer prevention and treatment. Prog Biophys Mol Biol 92:49–59

Houghton LA, Vieth R 2006 The case against ergocalciferol (vitamin D2) as a vitamin supplement. Am J Clin Nutr 84:694–697

Vieth R, Walfish PG 2000 How much vitamin D supplementation do adults require to ensure a serum 25(OH)D optimal for suppression of PTH? J Bone Miner Res 15:suppl 1:S231

Wharton BA, Darke SJ 1982 Infantile hypercalcaemia. In: Jelliffe EFP, Jelliffe DB (eds) Adverse effects of foods. Plenum p 397–404

Selenium

Jan Alexander

Department of Food Safety and Nutrition, Norwegian Institute of Public Health (NIPH), and Institute for Cancer Research and Molecular Medicine, Norwegian University of Science and Technology

Abstract. Selenium occurs as inorganic selenite or selenate and in organic forms in plants and other organisms used for food. The human selenoproteome consists of 25 selenoproteins. The main groups are glutathione peroxidases 1–5, iodothyronine deiodinases 1–3, thioredoxin reductases, selenoprotein P (SelP), and other proteins mostly with unknown function. In selenoproteins selenium occurs as selenocysteine. SelP works as a transporter of selenium between the liver and other organs. Selenium in the form of selenomethionine can also unspecifically substitute for methionine in other proteins. No specific deficiency condition has been described in humans. The aetiology of Keshan disease, a cardiomyopathy, is a combination of coxsackie virus and low selenium. Selenium status has been linked to the incidence of cancer and other diseases. Excess selenium can produce selenosis in humans affecting liver, skin, nails and hair. Recommended intake and upper tolerable level are 40–55 and 300 μg/day. A better chemical characterization of selenium compounds in foods and in particular supplements as well as knowledge on the apparent differences in biological activity between selenium compounds, both with respect to nutrition, disease protection and adverse effects, are needed. Supplementation studies should in addition to possible beneficial effects also focus on the possibility of possible adverse effects.

2007 Dietary supplements and health. Wiley, Chichester (Novartis Foundation Symposium 282) p 143–153

Selenium enters the food chain through plants and because of its great variation in soil the selenium contents in food vary geographically, with areas of selenium deficiency and excess. Humans are exposed to a variety of chemical forms of selenium: as inorganic salts, mainly selenate or selenite, and a large number of organic bound species produced in plants and other living organisms used as food—selenomethionine, selenocysteine, various methylated forms, derivatives of glutathione and selenosugars (Hogberg & Alexander 2007, Suzuki et al 2005, Whanger 2004).

Selenium as a nutrient

Selenium was established as an essential trace element 50 years ago. Since then it has been revealed that the human selenoproteome consists of 25 selenoproteins

143

serving several important biochemical functions (Gromer et al 2005, Kryukov et al 2003). In addition to five glutathione peroxidases (GSHPx1–5), which protect the organism against oxidative damage by reducing lipoperoxides and hydrogen peroxide, a second group is thioredoxin reductases that catalyse reduction of oxidized cellular proteins and play a role in redox regulation and resistance to oxidative stress and apoptosis (Gromer et al 2004). A third group is iodothyrodine deiodinases, which control the level of active triiodothyronine. Selenoprotein P in plasma probably has a transport function for selenium between the liver and other organs such as the brain and testes (Burk et al 2006a, Burk & Hill 2005, Hill et al 2007). Other selenoproteins seem to have structural functions in sperm and muscles. Lastly, there are several other selenoproteins mostly with unknown function. Most selenoproteins contain one selenium residue as selenocysteine per chain, whereas selenoprotein P contains 10 residues. The selenol group has a lower pK_a than the corresponding thiol group in cysteine and is less sensitive towards changes in pH (Hogberg & Alexander 2007).

Selenium from foods, either as inorganic salts or organic compounds, is converted through several steps to selenide, from which selenocysteine is formed via selenophosphate and serine on its tRNA. Selenocysteine is then incorporated into polypeptides guided by the UGA codon in all selenoprotein mRNAs and a selenocysteine insertion element in the non-coding region (Gromer et al 2005). Disruption of the gene for selenocysteine tRNA is embryonically lethal (Hogberg & Alexander 2007).

In excess, selenide may undergo redox cycling and cause toxicity (Anundi et al 1984). Selenide eventually is converted to excretable methylated forms or selenosugars (e.g. methyl 2-acetamido-2-deoxy-1-seleno-β-D-galactopyranoside), the latter being a major metabolite in human urine (Kuehnelt et al 2005, 2006).

Studies on experimentally induced selenium deficiency indicate hierarchies for selenium incorporation into selenoproteins. Methylation of selenocysteine tRNA, being dependent on selenium status, appears to play a role in regulation of the mammalian selenocysteine insertion machinery (Gromer et al 2005, Hogberg & Alexander 2007). A corresponding hierarchy also exists in humans exemplified by the fact that plasma GSHPx responded prior to plasma selenoprotein P in humans from a selenium-deficient region undergoing selenium supplementation (Xia et al 2005).

While all 25 selenoproteins contain selenocysteine, dietary selenomethionine at the expense of methionine may be unspecifically incorporated into many proteins (Hogberg & Alexander 2007). This is clearly observed in humans replete in selenium, as determined by maximum levels of plasma GSHPx and selenoprotein P, undergoing supplementation with selenomethionine or selenite. An increase in plasma selenium is only seen with selenomethionine pure, or from selenium yeast (Burk et al 2006b).

Selenium deficiency and diseases associated with selenium in human

Well-defined deficiency diseases have been described in many farm animals and laboratory animals, and include dysfunction of several organs and growth retardation (Diplock 1976). No specific deficiency condition has been described in humans, but several diseases have been associated with low dietary selenium (Alexander & Meltzer 1995, Hogberg & Alexander 2007, Rayman 2000, 2002).

Keshan disease is an endemic cardiomyopathy in children and pregnant women who live in the Keshan region of China, where selenium levels in food are extremely low. The incidence of the disease was dramatically reduced upon supplementation with sodium selenite (Ge & Yang 1993). However, a cardiotoxic coxsackie virus has also been implicated as part of the aetiology since enteroviruses, and in particular coxsackie viruses, were isolated from Keshan disease victims (Beck et al 2003). An amyocarditic strain of coxsackie virus B3,CVB3/0 converted to virulence when inoculated into selenium-deficient mice. This did also happen in selenium-replete GSHPx-1 knockout mice, but not in normal selenium-replete mice.

A large number of experiments, probably some hundreds, have been carried out to assess the effect of selenium compounds in various animal tumour models (Whanger 2004). In general the majority show some inhibitory effect on a variety of chemically induced cancers in skin, liver, colon, and breast as well as spontaneous mammary and intestinal tumours. Most experiments have used inorganic forms of selenium or organic selenium from yeast or garlic, e.g. methylselenocysteine, but various synthetic forms, e.g. ebselen, benzyl selenocyanate, have also been shown to be effective. It appears that the various compounds differ with respect both to anticarcinogenic potency and ability to support synthesis of selenoenzymes. It is well recognized that the most effective doses of selenium for anticarcinogenesis are above nutritional or at pharmacological levels. Apparently selenium works best in early phases of carcinogenesis and manifest tumours respond poorly. Several mechanisms of action have been proposed, but none have been proven (Rayman 2005, Whanger 2004).

A number of epidemiological studies with different designs have been carried out. These include the early ecological studies, which showed that geographical areas with high selenium levels (as measured in blood or food) had lower incidence of breast, digestive and lung cancer as well as lymphomas (Rayman 2005, Whanger 2004). Five sets of intervention trials from China have given support to efficacy of selenium against liver cancer risk, and oesophageal and stomach cancer, and lung cancer (Rayman 2005, Whanger 2004).

In a randomized double-blind placebo controlled trial in USA, older patients with a history of basal and/or squamous cell carcinoma were given 200 µg Se/day as selenium-enriched yeast or placebo (Clark et al 1996). The study started in 1983 and was analysed in 1993. Selenium treatment did not affect recurrent skin cancers, but reduced levels of non-skin cancer, including lung cancer, colorectal cancer and prostate cancer were observed. In a recent report following the patients to 1996, selenium supplementation was associated with an increased risk of squamous cell carcinomas and total non-melanoma skin cancer (Duffield-Lillico et al 2003a, b). The effect of antioxidant supplementation, including selenium, was evaluated for gastro-intestinal cancers in a systematic review and meta-analysis (Bjelakovic et al 2004). In four trials, of which three had unclear or inadequate methodology, selenium had significant beneficial effect on the incidence of gastro-intestinal cancer. Currently there are more trials on-going or planned (Rayman 2005).

Several other diseases are associated with low levels of selenium in serum or blood. Patients with liver cirrhosis, especially alcoholic cirrhosis, had very low levels of selenium in both serum and liver (Aaseth et al 1982, Hogberg & Alexander 2007). Some improvement regarding morning stiffness was observed in a double-blind study in patients with rheumatoid arthritis supplemented with selenium and vitamin E (Aaseth et al 1998). Combined selenium and iodine deficiency may result in sever myxoedematous cretinism (Vanderpas et al 1990). Recently low selenium status, as assessed by toe nail selenium, was associated with a 4.4 times increased risk of preeclampsia in a case control study in women from the UK (Rayman et al 2003).

Nutritional requirement and recommendations

Several approaches were used to establish the nutritional requirement. Studies from China and elsewhere indicated that an intake of $20\,\mu g/day$ was not associated with Keshans disease (Alexander et al 2004). Variable results have been obtained using 'saturation' of plasma selenoproteins. These estimations have also been complicated by the fact that the studies used different forms of selenium with different measures of selenium status. In a supplemental study from China (Xia et al 2005) maximum enzyme activity of plasma GSHPx was reached with a supplement of $37\,\mu g/day$ of selenomethionine whereas $66\,\mu g/day$ of selenite was required to reach the same maximum. In a study from New Zealand (Duffield et al 1999) an upper estimated requirement of $90\,\mu g/day$ was necessary to reach maximum plasma GSHPx activity. The physiological relevance of the 'saturation of selenium-dependent enzymes approach' can be questioned. The average requirements for adults have by different bodies been set between 30 and $45\,\mu g/day$, whereas the recommended intakes for adults have been set between 40 and $55\,\mu g/day$ (Alexander et al 2004, EU SCF 2000). The current intake in Europe is generally low in comparison to USA, Japan and Canada (Rayman 2005).

Selenium and adverse effects

Adverse effects of high intakes of selenium have been reported from regions in China with soil naturally high in selenium. The syndrome of selenosis comprises the following symptoms: brittle hair, nail changes (thickening, streaks, paronychia), skin changes on extensor sides of arms, hyper-reflexia and liver effects. At high doses, neurotoxicity with convulsions and paralysis may occur (EU SCF 2000, Hogberg & Alexander 2007, Yang et al 1989a, b, 1983).

Safety assessment—tolerable upper level

For establishing an upper level of selenium intake the EU Scientific Committee on Food used the Chinese study on selenosis and liver toxicity following high dietary intake from rice and maize of selenium mainly in the form of selenomethionine (EU SCF 2000). A lowest observable adverse effect level (LOAEL) and a no observable adverse effect level (NOAEL) for clinical selenosis at 900–1000 µg/day and 850 µg/day was derived, and using an uncertainty of factor of three for remaining uncertainty an upper level (UL) of 300 µg/day was established. In support of this UL was the long term supplemental study of Clarke et al (1996), which did not record any adverse effects, particularly selenosis or liver toxicity at an intake of 290 µg/day. The UL applies to selenium from food, usually selenomethionine and supplements approved for use, e.g. inorganic compounds.

It has been claimed that the UL is set too low and in support it is referred to a number of often small supplemental studies on possible beneficial effects. A major drawback of these studies is that they have not been designed for the detection of possible adverse. Furthermore, selenium supplements are most often very poorly characterized chemically and it is known that they may contain a lot of different selenium compounds.

Conclusion

Selenium research during 50 years from the early findings of a role in the scavenging of peroxides to the present understanding of its many and widely different biological functions, is a fascinating story.

The basic nutritional needs to avoid deficiency seem to be fairly well established, whereas the documentation for beneficial effects of additional intake of selenium with regard to many diseases, in particular cancer, is insufficient. Still the biological functions of many selenoproteins are unknown. This knowledge is pivotal for disease prevention. Current studies do not differentiate between nutritional effects and pharmacological effects occurring after higher doses of selenium. Such distinctions should be made. There is a need for better chemical characterization of selenium compounds in foods and in particular supplements. Also the apparent differences in biological activity, both with respect to nutrition, disease protection and adverse effects, between selenium compounds need to be examined. Supplementation studies should, in addition to possible beneficial effects, focus on the possibility of possible adverse effects.

References

Aaseth J, Alexander J, Thomassen Y, Blomhoff JP, Skrede S 1982 Serum selenium levels in liver diseases. Clin Biochem 15:281–283

Aaseth J, Haugen M, Forre O 1998 Rheumatoid arthritis and metal compounds—perspectives on the role of oxygen radical detoxification. Analyst 123:3–6

Alexander J, Meltzer HM 1995 Selenium. In: Oskarsson A (ed) Risk Evaluation of Essential Trace Elements - essential versus toxic levels of intake. Copenhagen: Nordic Council of Ministers, Nord 18:15–65

Alexander J, Andersen SA, Aro A et al 2004 Nordic Nutrition Recommendations 2004. Copenhagen: Nordisk Ministerråd, Nord, 13

Anundi I, Stahl A, Hogberg J 1984 Effects of selenite on O2 consumption, glutathione oxidation and NADPH levels in isolated hepatocytes and the role of redox changes in selenite toxicity. Chem Biol Interact 50:277–288

Beck MA, Levander OA, Handy J 2003 Selenium deficiency and viral infection. J Nutr 133:1463S–1467S

Bjelakovic G, Nikolova D, Simonetti RG, Gluud C 2004 Antioxidant supplements for prevention of gastrointestinal cancers: a systematic review and meta-analysis. Lancet 364:1219–1228

Burk RF, Hill KE 2005 Selenoprotein P: an extracellular protein with unique physical characteristics and a role in selenium homeostasis. Annu Rev Nutr 25:215–235

Burk RF, Hill KE, Motley AK, Austin LM, Norsworthy BK 2006a Deletion of selenoprotein P upregulates urinary selenium excretion and depresses whole-body selenium content. Biochim Biophys Acta 1760:1789–1793

Burk RF, Norsworthy BK, Hill KE, Motley AK, Byrne DW 2006b Effects of chemical form of selenium on plasma biomarkers in a high-dose human supplementation trial. Cancer Epidemiol Biomarkers Prev 15:804–810

Clark LC, Combs GF Jr, Turnbull BW et al 1996 Effects of selenium supplementation for cancer prevention in patients with carcinoma of the skin. A randomized controlled trial. Nutritional Prevention of Cancer Study Group. JAMA 276:1957–1963

Diplock AT 1976 Metabolic aspects of selenium action and toxicity. CRC Crit Rev Toxicol 4:271–329

Duffield AJ, Thomson CD, Hill KE, Williams S 1999 An estimation of selenium requirements for New Zealanders. Am J Clin Nutr 70:896–903

Duffield-Lillico AJ, Dalkin BL, Reid ME et al 2003a Selenium supplementation, baseline plasma selenium status and incidence of prostate cancer: an analysis of the complete treatment period of the Nutritional Prevention of Cancer Trial. BJU Int 91:608–612

Duffield-Lillico AJ, Slate EH, Reid ME et al 2003b Selenium supplementation and secondary prevention of nonmelanoma skin cancer in a randomized trial. J Natl Cancer Inst 95:1477–1481

EU SCF 2000 EU Scientific Committee on Food: Opinion on Selenium Expressed on 19 October 2000. European Commission

Ge K, Yang G 1993 The epidemiology of selenium deficiency in the etiological study of endemic diseases in China. Am J Clin Nutr 57:259S–263S

Gromer S, Urig S, Becker K 2004 The thioredoxin system—from science to clinic. Med Res Rev 24:40–89

Gromer S, Eubel JK, Lee BL, Jacob J 2005 Human selenoproteins at a glance. Cell Mol Life Sci 62:2414–2437

Hill KE, Zhou J, Austin LM et al 2007 The selenium-rich C-terminal domain of mouse selenoprotein P is necessary for supply of selenium to brain and testis but not for maintenance of whole-body selenium. J Biol Chem 282:10972–10980

Hogberg J, Alexander J 2007 Selenium. In: Nordberg GF, Fowler B, Nordberg M, Friberg L (eds) Handbook on the Toxicology of Metals. New York, Amsterdam: Academic Press/Elsevier

Kryukov GV, Castellano S, Novoselov SV et al 2003 Characterization of mammalian selenoproteomes. Science 300:143–1443

Kuehnelt D, Kienzl N, Traar P, Le NH, Francesconi KA, Ochi T 2005 Selenium metabolites in human urine after ingestion of selenite, L-selenomethionine, or DL-selenomethionine: a quantitative case study by HPLC/ICPMS. Anal Bioanal Chem 383:235–246

Kuehnelt D, Juresa D, Kienzl N, Francesconi KA 2006 Marked individual variability in the levels of trimethylselenonium ion in human urine determined by HPLC/ICPMS and HPLC/vapor generation/ICPMS. Anal Bioanal Chem 386:2207–2212

Rayman MP 2000 The importance of selenium to human health. Lancet 356:233–241

Rayman MP 2002 The argument for increasing selenium intake. Proc Nutr Soc 61:203–215

Rayman MP 2005 Selenium in cancer prevention: a review of the evidence and mechanism of action. Proc Nutr Soc 64:527–542

Rayman MP, Bode P, Redman CW 2003 Low selenium status is associated with the occurrence of the pregnancy disease preeclampsia in women from the United Kingdom. Am J Obstet Gynecol 189:1343–1349

Suzuki KT, Kurasaki K, Okazaki N, Ogra Y 2005 Selenosugar and trimethylselenonium among urinary Se metabolites: dose- and age-related changes. Toxicol Appl Pharmacol 206:1–8

Vanderpas JB, Contempre B, Duale NL et al 1990 Iodine and selenium deficiency associated with cretinism in northern Zaire. Am J Clin Nutr 52:1087–1093

Whanger PD 2004 Selenium and its relationship to cancer: an update dagger. Br J Nutr 91:11–28

Xia Y, Hill KE, Byrne DW, Xu J, Burk RF 2005 Effectiveness of selenium supplements in a low-selenium area of China. Am J Clin Nutr 81:829–834

Yang GQ, Wang SZ, Zhou RH, Sun SZ 1983 Endemic selenium intoxication of humans in China. Am J Clin Nutr 37:872–881

Yang G, Yin S, Zhou R et al 1989a Studies of safe maximal daily dietary Se-intake in a seleniferous area in China. Part II: Relation between Se-intake and the manifestation of clinical signs and certain biochemical alterations in blood and urine. J Trace Elem Electrolytes Health Dis 3:123–130

Yang G, Zhou R, Yin S et al 1989b Studies of safe maximal daily dietary selenium intake in a seleniferous area in China. I. Selenium intake and tissue selenium levels of the inhabitants. J Trace Elem Electrolytes Health Dis 3:77–87

DISCUSSION

Rayman: You didn't mention the nice work on immune effects by Malcolm Jackson's group at Liverpool (Broome et al 2004). They have looked at people in the UK, giving them either 50 or 100 µg selenium, and demonstrating significant beneficial effects on immune function. This also enabled these people to clear attenuated polio virus more rapidly and with fewer mutations in the viral genome. This brings me on to what Jan Alexander was saying about the work being done on selenium and various viruses. In selenium-deficient individuals or animals the virus mutates. These are RNA viruses so they have no repair ability. The published mechanism seems to be that glutathione peroxidase is preventing oxidative damage to the viral genome. So when the host is deficient in selenium there is an increased virulence and worse outcome. Moving on to the anticancer effects, which are probably highest for prostate cancer, Jan mentioned that the mechanism wasn't understood. There are many mechanisms discussed, but recently a paper was published showing that the anticancer effects can be linked to two aspects. One is the production of low molecular weight selenium compounds such as methylselenol, which can be produced from selenomethyl selenocysteine, which is a component of some foods and also of selenium yeast. We also know that selenoproteins are involved. The link between selenoprotein polymorphisms and cancer risk demonstrates this. Lastly, Raymond Burk has recently suggested that the American RDA ($55\,\mu g/day$) is now not sufficiently high and needs to be revised upwards, because it doesn't show optimization of all selenoproteins (Xia et al 2005). As far as the toxicity is concerned, there have been some much longer studies using doses up to $1600\,\mu g/day$ in men with prostate cancer which haven't shown adverse effects (Reid et al 2004). In the Clark et al (1996) study they now have more than 10 years of data on people taking $200\,\mu g/day$ with no more adverse effects than in the placebo group.

Azzi: Why does glutathione peroxidase, which eliminates hydrogen peroxide, increase viral clearance?

Rayman: The viral genome becomes damaged.

Azzi: The problem is that it eliminates hydrogen peroxide, and it is hydrogen peroxide that is producing damage to the virus, together with metal species which are able to produce OH radicals. Another hypothesis is that the virus is not mutated by the lack of selenium, and that under conditions of lack or diminution of selenium, the virulent virus species become selected.

Rayman: But iron has the opposite effect. If you use iron in those same systems with the same virus, you get the opposite effect.

Stocker: You mentioned that in animals such as the pig, cardiomyopathy can be demonstrated. That's a well established model. Is there any evidence that in pigs, like in humans, there may be a link with viral infection in addition to selenium deficiency?

Alexander: I don't think so. These mechanistic considerations about myopathies as a result of deficiency in various species implicate various selenoproteins in muscles. However these are not established mechanisms but more on a speculative basis for the moment. I would say the same applies to the cancer mechanism. There are a lot of suggestions, but we really don't know.

Halliwell: We know with vitamin E that you can probably get enough from diet, and deficiency is rare. With folic acid we need some kind of supplement to reach optimal intake. The same may be true for vitamin D and Ca^{2+}, particularly if you don't go out in the sun. So what is the situation for selenium? Can we get optimum intake from diet?

Alexander: We don't really know what the optimum intake is. I don't see that there's any reason that we need 600 µg/day. Is it 100, or 120 or 50 µg? It's difficult to determine.

Halliwell: You showed how blood selenium in several UK cities is dropping because we eat less wheat from the USA. Does this matter?

Rayman: I think it does. There is some evidence, and one of the strongest pieces of evidence is about the effect of selenium on the immune system. I referred to the work of Malcolm Jackson's group in Liverpool that does show that our immune function could be improved by a small amount of selenium (Broome et al 2004). He found that even the group that took 100 µg selenium in addition to their background intake (30 µg) needed to go higher than this. The Food Standards Agency has now funded further studies to look at this. When these people were given an attenuated polio virus they didn't clear it very efficiently and there were a lot of viral mutations. There is ever increasing and ever stronger evidence on various cancers. There have been some excellent studies on cancers of liver, lung and pancreas, as well as a bevy of studies on prostate cancer. There is also an intervention in prostate cancer underway in the USA. My worry about this is that in some of the individuals in that study, because selenium status is so much higher in the USA, it isn't impossible that the threshold of benefit will be exceeded. It would have been better to do this study in Europe where selenium status is considerably lower.

Aggett: Can you think of a strategy for determining what an optimum selenium intake would be? Should we address getting a full complement of the selenoproteins that you showed as a reference, or are there any dose–response data to inform our understanding of the metabolic chain you developed?

Alexander: I'm not aware of a useful dose–response. We don't know what the end point is. What is the beneficial effect we should look for? All these studies on cancer indicate convincingly that it has some protective effect, but it is difficult to draw quantitative information from those studies.

Aggett: So we just don't have the basic fundamental information. We are happy looking at the very low and high levels, but what about somewhere in between?

Alexander: We could use plasma glutathione peroxidase or selenoprotein P as endpoint markers, but we are not sure about their possible relationship with disease endpoints.

Aggett: I got the impression that glutathione peroxidase indicated an intake of about 1 µg/kg body weight. We are falling pretty short of that.

Rayman: In the UK we are about half of that.

Przyrembel: There are some small studies in selenium-deficient PKU children which have been performed in a stepwise fashion with selenium yeast. Unfortunately, these studies were stopped after the selenium blood levels didn't show that they would arrive at any plateau, whereas glutathione peroxidase activity reached a plateau. The researchers got frightened and stopped the study. It also depends on the judgment of the researchers about the endpoint. Do you want 100% glutathione peroxidase activity, or will 75% do?

Rayman: But we also need to look at the disease endpoint. The amount of selenium needed will differ for each disease. Some will not necessarily be related to selenoproteins.

Alexander: Yes, for example, the effect on cancer could be due to low molecular weight compounds and not specifically to the proteins. I am not sure that we should aim at the full expression of all the selenoproteins. It doesn't seem very logical.

Rayman: In the USA population they do have full expression of the selenoproteins.

Cashman: You have to be careful in some of the intervention/supplementation trials. With Ca^{2+} we don't usually have a dose–response design within our trials, due to the many constraints that would involve. You might increase intake by 800 mg and see a benefit, but the real increment that you need over habitual intakes might be 200 mg. This may be achievable by public health means.

Rayman: It is a shame that in the select trial underway in the USA that they didn't use a dose of 100 as well as a dose of 200. At 200 we will be talking about a supplement.

Alexander: I agree that these studies should be carried out in Europe, not in the USA.

Cashman: Our lack of biomarkers is another knowledge gap. If we had them we could do the dose–response studies to inform the design of trials using harder outcome measures, e.g. bone mineral density.

Halliwell: Am I correct in thinking that selenoprotein P is a selenium transporter?

Rayman: That's one of its main functions. It is thought to have some specific functions in the brain, but it does carry around 10 selenium residues. It is also thought to be a peroxinitrite scavenger.

References

Broome C, McArdle F, Kyle J et al 2004 An increase in selenium intake improves immune function and poliovirus handling in adults with marginal selenium status. Am J Clin Nutr 80:154–162

Clark LC, Combs GF Jr, Turnbull BW et al 1996 Effects of selenium supplementation for cancer prevention in patients with carcinoma of the skin. A randomized controlled trial. Nutritional Prevention of Cancer Study Group. JAMA 276:1957–1963 (erratum: 1997 JAMA 277:1520)

Reid ME, Stratton MS, Lillico AJ et al 2004 A report of high-dose selenium supplementation: response and toxicities. J Trace Elem Med Biol 18:69–74

Xia Y, Hill K, Byrne D, Xu J, Burk R 2005 Effectiveness of selenium supplements in a low-selenium area of China. Am J Clin Nutr 81:829–834

Herbal medicines: balancing benefits and risks

Edzard Ernst

Complementary Medicine, Peninsula Medical School, Universities of Exeter & Plymouth, 25 Victoria Park Road, Exeter EX2 4NT, UK

Abstract. Herbal medicines are preparations containing exclusively plant material. Their efficacy can be tested in clinical trials much like synthetic drugs but numerous methodological and logistical problems exist. For several herbal medicines, efficacy has been established; for many others, this is not the case mostly because the research has not been done. Many consumers believe that herbal medicines are natural and therefore safe. This is a dangerous simplification. Some herbal medicines are associated with toxicity others interact with synthetic drugs. The often under-regulated quality of herbal medicines amounts to another safety issue. Contamination or adulteration of herbal medicines are possible and can cause harm. In order to conduct a risk-benefit analysis of a specific herbal medicine for a specific indication, we require definitive efficacy and safety data. This is currently the case for only very few such preparations. It follows that, in order to advise consumers responsibly, the gaps in our present knowledge require filling.

2007 Dietary supplements and health. Wiley, Chichester (Novartis Foundation Symposium 282) p 154–172

Herbal medicine (HM) has many definitions; in this chapter I refer to it as the medicinal use of preparations that contain exclusively plant material (Ernst et al 2006). There are, of course, many forms of HM. Traditional herbalists (e.g. European traditional herbalism, traditional Chinese medicine, Kampo medicine) usually individualize prescriptions containing several herbs according to the patient's symptoms, constitution, etc. They also employ diagnostic principles that are considered obsolete in conventional medicine. The evidence for this traditional approach is scant and far from convincing.

In recent decades an entirely different approach has emerged, sometimes termed 'rational phytotherapy' (Schulz et al 2001). It views HMs as complex drugs that are understood in terms of pharmacological principles. This means that one HM is normally used for one medical condition (diagnosed in the conventional way) for which it has been shown to be effective in clinical trials. It is this type of HM that is addressed in this chapter.

Problems with clinical research in herbal medicine

Clinical trials are prospective experiments for assessing the nature and results of medical interventions. Generally speaking, clinical trials of HM should follow the same lines as any other such experiment (Ioannidis et al 2004). In particular, they should provide clear definitions or descriptions of:

- Research questions/hypothesis
- Participants and sample size
- Intervention
- Outcome measurements
- Randomisation procedures
- Blinding
- Statistical analysis
- Results

But there are nevertheless many differences between HM and synthetic drugs (Table 1). One significant logistical problem in relation to clinical trials of HM is the lack of funds for such research. There is little patent protection for medicinal plants; thus little impetus exists for initiating expensive clinical trials. Moreover, herbal manufacturers are usually relatively small companies. Lack of funds leads to a paucity of research and the regrettable fact that studies are often not as rigorous as they could be. This is aggravated by a lack of research culture and expertise in HM. If one considers that the clinical effects of HM are usually mild and take a relatively long time to become clinically manifest (Ernst et al 2006), it becomes clear that large, long-term studies are often required in order to demonstrate efficacy—a fact which can significantly complicate the existing logistical problems with trials of HM. In the USA, the NIH is now supporting large

TABLE 1   Some of the differences between herbal medicines and synthetic drugs

Herbal medicine	Synthetic drug
More than one pharmacologically active compound	Usually one pharmacologically active compound
Active ingredients often not known	Active ingredient known
Pure compound not available	Pure compound available
Raw material limited	Raw material unlimited
Quality variable	Quality consistent
Mechanism of action often unknown	Mechanism of action known
Toxicology often unknown	Toxicology known
Long tradition of use	Short tradition of use
Therapeutic window wide	Therapeutic window narrow
Adverse effects normally rare	Adverse effects of many drugs frequent

clinical trials of HM. Other countries are less fortunate, and globally HM research remains grossly under-funded.

In addition to these general difficulties, clinical trials of HM are confronted with a range of problems that are not specific to this area but, in terms of their importance, represent obstacles to research.

Bias

Bias is, of course, an issue in any research; in HM it is probably larger than elsewhere. We have shown that countries such as Germany (Giovannini et al 2004) or China (Tang et al 1999), which are of obvious importance to HM, publish predominantly positive results. This does certainly not prove the existence of bias, it merely suggests it. Similarly, complementary medicine journals are associated with a strong positive publication bias (Ernst & Pittler 1997). The concern therefore is that, perhaps due to the above-mentioned lack of research culture in this area, bias is more of an issue in herbal than in conventional medicine (Ernst & Pittler 1997).

Outcome measures

Outcome measures used in clinical trials should, of course, be validated. In trials of HM, soft and non-validated outcome measures are often employed, e.g. percentage of patients perceiving benefit or patients' preference. Similarly, multiple outcomes are frequently used without accounting for multiple statistical tests. Finally, surrogate endpoints are frequent and researchers sometimes seem to measure what is measurable rather than what is relevant.

Blinding

Blinding can be a real problem in 'double-blind' trials of HM. Due to taste, odour or appearance, HMs may be distinguishable from placebos. Thus unblinding can occur and influence the results. For instance, there are numerous 'placebo-controlled, double-blind' trials of garlic preparations for cholesterol lowering (Stevinson et al 2000). Yet anyone who has ever conducted such a study knows that, due to the body odour caused by regular garlic intake, blinding is not a realistic option.

The intervention

The intervention must, of course, be fully described in all clinical trials. The aim must be to disclose all details such that the study can be reproduced elsewhere. In HM, this can be more complex than in conventional studies. Herbal medicines are natural products; their composition and (therefore effects) could depend on a range of factors, for example:

- Source(s), e.g. soil, climate
- Processing/extraction
- Storage.

A degree of variability from batch to batch is often unavoidable. Additional issues can be adulteration (Ernst 2002a) and contamination, e.g. with heavy metals (Ernst et al 2001).

An often-voiced criticism of clinical trials is that such experiments tell us little about individual patients. In HM, treatments are often highly individualized. Therefore, some proponents dismiss the value of clinical trials of HM. This obviously ignores the existence of n of 1 studies. Such trials are clearly possible in HM and they inform us about responses of individual patients (Canter & Ernst 2003). Similarly, numerous modifications of the standard design of clinical trials have been developed (Ernst 2004a), some of which may be suitable for specific research questions encountered in HM.

Despite these problems, clinical trials of HM are often of surprisingly good methodological quality. In fact, a comparison of RCTs of conventional and complementary (including HM) therapies recently suggested that their methodological quality was, on average, similar: 48% of the former and 54% of the latter were rated as of high quality (Klassen et al 2005).

Systematic reviews and meta-analyses (systematic reviews that include statistical pooling of primary data) are projects evaluating the totality of the available evidence. These approaches minimize random and selection biases and thus provide a more reliable tool for healthcare decision making (Egger et al 2001). Numerous systematic reviews of HM have become available during the last decade (my own team has published about 50 such articles; a full list is available from the authors). A comparison of systematic reviews of conventional and complementary (including HM) therapies showed that the average methodological quality of the latter was superior to that of the former (Lawson et al 2005).

Efficacy

Given the fact that there are hundreds, if not thousands of HMs, it is nonsensical to even try to define the efficacy of HM as such. This multitude also renders a comprehensive review of the efficacy of single HMs difficult. My team recently published an evidence-based overview of this area (Ernst et al 2006). For this purpose we classified the 'weight' of the evidence separately from the direction of the evidence. The 'weight' was defined as a composite measure of three independent variables:

- the level of the available evidence (e.g. single clinical trials or an up-to-date systematic review)
- the methodological quantity of that evidence
- its quantity (i.e. total sample size).

The direction of the evidence was categorized as 'clearly positive', if the bulk of the available data demonstrated efficacy of the HM in question. It was 'clearly negative' when all the available evidence suggested lack of efficacy. In many instances, the primary data was neither clearly positive nor negative, in such situations it was rated as either 'tentatively positive' or 'tentatively negative' or 'uncertain'.

Using this approach, we identified several HMs with proven efficacy, i.e. maximal 'weight' of evidence and direction clearly positive (Table 2). In several other instances the evidence was clearly positive but its 'weight' was not rated as maximal; these are HMs which are supported by encouraging albeit not fully convincing evidence (Table 3). For several further HMs we found that the evidence was clearly negative (but in all cases was the 'weight' maximal) (Table 4).

TABLE 2   Herbal medicines which have been demonstrated in clinical trials to be effective[a] for defined conditions

Herbal Medicine	Condition
African plum	Benign prostatic hyperplasia
Allium vegetables	Cancer prevention
Ephedra sinica	Overweight/obesity
Ginkgo biloba	Alzheimer's disease
Ginkgo biloba	Peripheral arterial occlusive disease
Green tea	Cancer prevention
Guar gum	Diabetes
Guar gum	Hypercholesterolemia
Hawthorn	Chronic heart failure
Horse chestnut	Chronic venous insufficiency
Kava[b]	Anxiety
Oat	Hypercholesterolemia
Padma 28	Peripheral arterial occlusive disease
Peppermint and caraway	Non-ulcer dyspepsia
Phytodolor	Osteoarthritis
Phytodolor	Rheumatoid arthritis
Psyllium	Constipation
Psyllium	Diabetes
Red clover	Menopause
Sal palmetto	Benign prostatic hyperplasia
St John's wort	Depression
Soy	Hypercholesterolemia
Tomato (Lycopene)	Cancer prevention
Yohimbine[b]	Erectile dysfunction

[a] Maximum 'weight' of evidence and direction of result clearly positive. Data extracted from Ernst et al (2006).
[b] Risk may not outweigh the benefit.

TABLE 3 Herbal medicines which, according to clinical trials, are probably effective[a] for defined conditions

Herbal medicine	Condition
Achillea wilhelmsii	Hypertension
Andrographis paniculata	Upper respiratory tract infection
Artichoke	Non-ulcer dyspepsia
Astragalus	Chronic heart failure
Boswellia serrata	Osteoarthritis
Butterbur	Migraine
Calendula	Cancer palliation
Capsaicin	Osteoarthritis
Clinacanthus nutans	Herpes zoster
Curcuma domestica	Non-ulcer dyspepsia
Fenugreek	Diabetes
Fenugreek	Hypercholesterolemia
Florelax	Irritable bowel syndrome
Garlic	Upper respiratory tract infection
Ginger	Nausea, motion sickness
Ginger	Nausea of pregnancy
Ginseng	Cancer prevention
Ginseng (prevention)	Upper respiratory tract infection
Gotu kola	Chronic venous insufficiency
Hibiscus	Hypertension
Ipomoea batatas	Diabetes
Kava	Insomnia
Kava	Menopause
Konjac glucomannan	Hypercholesterolemia
Korean red ginseng	Erectile dysfunction
Lemon balm (herpes labialis, treatment)	Herpes simplex
Maritime pine	Hypertension
Melissa officinalis	Alzheimer's disease
Nettle	Benign prostatic hyperplasia
Padma Lax	Constipation
Padma Lax	Irritable bowel syndrome
Peppermint oil (local)	Headache
Peppermint	Nausea and vomiting postoperative
Pomegranate	Hypertension
Red clover	Diabetes
Red pepper	Non-ulcer dyspepsia
Salvia officinalis	Alzheimer's disease
Siberian ginseng (prevention)	Herpes simplex
Stinging nettle	Osteoarthritis
Soy	Diabetes
Thunder god vine	Rheumatoid arthritis
Tinaspora cordifolia	Hay fever
Willow bark	Osteoarthritis

[a] Direction of result clearly positive but 'weight' of evidence not maximal. Data from Ernst et al (2006).

TABLE 4 Herbal medicines which have been tested in clinical trials and found to be ineffective (any 'weight' of evidence and direction of result clearly negative)

Herbal medicine	Condition
Aloe vera	Hypercholesterolemia
Capsaicin cream (symptomatic)	AIDS/HIV infection
Cranberry	Cancer palliation
Dong quai	Menopause
Echinacea purpurea (genital herpes)	Herpes simplex
Evening primrose oil	Menopause
Flaxseed oil (alpha-linoleic acid)	Rheumatoid arthritis
Ginkgo biloba	Depression
Ginkgo biloba	Drug/alcohol dependence
Grapeseed	Hay fever
Green algae	Hypertension
Guar gum	Overweight/obesity
Kudzu	Drug/alcohol dependence
Laetrile	Cancer 'cures'
LIV 52Q	Hepatitis
Olive leaf	Hypertension
Panax ginseng	Alzheimer's disease
Psyllium	Overweight/obesity
Pueraria lobata	Menopause
Red clover	Hypertension
Soy products	Cancer palliation
St John's wort	AIDS/HIV infection
Tea tree (herpes labialis, treatment)	Herpes simplex
Tinospora crispa	Diabetes
Uncaria gambir	Hepatitis
Valerian	Anxiety
Wild yam	Menopause
Xioke tea	Diabetes

Risks

The high prevalence of HM-usage renders rigorous safety assessments an ethical imperative. Yet such assessments can be complex and are often hampered by the incompleteness of the information. A more comprehensive analysis of this area is provided elsewhere (Ernst 2004b). For the purpose of this chapter, a few examples must suffice.

Toxicity: example of kava

Kava kava (*Piper methysticum* Forst.) is a perennial shrub native to the South Pacific where Kava drinking is part of traditional ceremonies and informal social occasions. In Europe, kava extracts became widely used for treating anxiety disorders because many clinical trials demonstrated that they are effective anxiolytics (Pittler & Ernst 2000)

Heavy and regular consumption of kava has been associated with poor health status including malnutrition and weight loss, liver and renal dysfunction, altered blood biochemistry and symptoms suggestive of pulmonary hypertension (Matthews et al 1998). A distinctive reversible ichthyotic rash known as kava dermopathy has been noted with heavy consumption of the beverage (Norton & Ruze 1994), and repeated episodes of generalised choreoathertosis secondary to kava bingeing have been reported in one individual (Spillane et al 1997). The amount of kava consumed in these cases is, however, considerably higher than recommended therapeutic doses.

In recommended therapeutic doses, kava has been associated only with a variety of mild and transient adverse effects such as erythema and neurological manifestations (Stevinson et al 2002). Several cases of liver damage occurring in association with ingestion of kava extracts have been reported (Stoller 2000). These included a 50-year-old man with jaundice and subsequent liver failure (Escher & Desmeules 2001) and five other patients who developed jaundice. Histological data from four patients were consistent with an immuno-allergic mechanism. Several of the affected patients had taken other medications with hepatotoxic potential. Symptoms generally occurred between three weeks and four months of starting kava extracts and involved doses containing 60 to 210 mg kavapyrones per day. The majority of cases involved acetone extracts, but one appeared to refer to an alcoholic extract of kavapyrones (Strahl 1998); this report also included a re-challenge: following normalization of liver values with discontinuation of all medications, a marked increase in transaminase levels occurred 14 days after this patient resumed use of kava. This increase was reversed upon discontinuation. Based on these suspicions (worldwide about 80 cases), kava has been taken off the market in several countries.

Interactions: example of St John's wort

In a recent systematic review, the safety profile of St John's wort (*Hypericum perforatum*), an HM with proven anti-depressive activity, was compared with those of several conventional antidepressants (Stevinson & Ernst 1999). St John's wort appeared to be better tolerated than synthetic antidepressants. Various case reports suggest that more serious adverse events of hypericum are possible including subacute toxic neuropathy (Bove 1998), psychotic relapse in schizophrenia (Lal & Iskandar 2000), delirium (Khawaja et al 1999), serotonin syndrome (Parker et al 2001), and hair loss (Parker et al 2001). Several incidents of hypomania have also been associated with St John's wort (Stevinson et al 2004).

Some of the drug interactions associated with modern antidepressants (Nemeroff et al 1996) also apply to hypericum. Evidence first emerged from case reports, and *in vivo* and pharmacokinetic studies indicating that hypericum affects the metabolism of several concomitant medications resulting in reduced plasma concentrations of the drug (Ernst 1999). Further experimental studies have confirmed that hypericum is a potent inducer of several cytochrome P450 enzymes (Roby et al 2000) and the transporter protein P-glycoprotein (Durr 2000). Both mechanisms work in concert to lower the plasma levels of xenobiotics, including about 50% of all prescription drugs. Pre-clinical studies have demonstrated the potential for interactions with digoxin (Johne et al 1999), indinavir (Piscitelli et al 2000), amitriptyline (Roots et al 2000) and other drugs. Cases of acute heart transplant rejections of patients on cyclosporine (Ernst 2002b) reduced anticoagulant effects of warfarin and intermenstrual bleeding with oral contraceptives (Yue et al 2000) have been reported.

Contamination: example of traditional Chinese medicines

Traditional Chinese medicines (TCMs) are treatments commonly advocated for a wide range of conditions (e.g. Ernst & White 2000, Eisenberg et al 1998). TCMs are usually complex mixtures of several (often 20 or more) different medicinal plants. They are usually prescribed by therapists or marketed as dietary supplements thus avoiding the usual quality standards. The toxicity of TCMs has been repeatedly reviewed (e.g. Ernst 2000, Chan & Critchley 1996), but other safety issues are often neglected. One such problem relates to contamination with heavy metals (Ernst 2000).

A systematic review (Ernst 2002a) summarized the recent literature relating to heavy metal contamination of TCMs. Twenty-two publications were included. Nine publications of case reports and five case series were found, comprising a total of 106 cases of heavy metal poisoning associated with TCM-use as a likely cause. For instance, the case of a 59-year old patient was reported, who had consumed around 15 mg of elemental lead daily through a TCM prescribed by a practitioner of Chinese medicine (Lightfoote et al 1977). Her signs and symptoms normalized with chelation therapy. In similar instances, the lead poisoning was largely asymptomatic and only discovered through routine screening (Chan et al 2001, Levitt et al 1984). In one Hong Kong infant, lead poisoning with TCM caused acute lead encephalopathy (Yu & Yeung 1987). Other US authors reported three cases of mercury poisoning associated with TCM use (Wu et al 1996a). Further case reports or case series refer to lead poisoning, (Markowitz et al 1994, Wu et al 1996a, b) cadmium poisoning (Wu et al 1996a), and arsenic poisoning (Tay & Seah 1975, Khawaja et al 1999). Parker et al (2001) screened 17 patients with cutaneous lesions related to chronic arsenicism in Taiwan. They found that 14 of them had a history of TCM intake. Eleven patients developed squamous cell carcinomas. A fatal case of arsenic poisoning through TCM was reported in 1998 from Hong Kong (Chi et al 1992), and Tay and Seah (1975) reported 74 cases of TCM-induced arsenic poisoning, 10 of which had malignancies at the time of investigation. An interesting case was recently described where a TCM was applied as a spray to treat mouth ulcers (Li et al 2000). The 5-year old patient developed motor and vocal tics as a result of the high mercury content of the spray. Two rare cases of combined thallium and lead poisoning were described by US authors (Schaumburg & Berger 1992). Both patients suffered from paresthesiae and dramatic hair loss.

Epidemiological investigations confirm that TCMs can be contaminated; 2803 Taiwanese subjects were randomly selected and screened for elevated blood lead levels (Chu et al 1998). A history of TCM-use was shown to be a significant risk factor for high blood levels with an odds ratio of 3.09 (95% CI: 1.60–5.97). Similarly, 319 Taiwanese children aged 1–7 years were screened for blood lead levels and their parents were interviewed about possible risk factors (Cheng et al 1998). The consumption of the TCM 'Ba-wa-san' was, in a multiple regression analysis, significantly ($P = 0.038$) associated with increased lead levels. Those children ($n = 66$) who had consumed 'Ba-wa-san' had a mean blood lead concentration of 4.96 µg/dL (SD = 2.71) and those who had not yielded a concentration of 4.13 µg/dL (SD = 2.83). This difference was statistically at the 1% level. Finally, several analytical studies confirm heavy metal contamination of TCMs (Myerson et al 1982, D'Alauro et al 1984, Chi et al 1992, Espinoza et al 1995, Koh & Woo 2000, Melchart et al 2001).

Adulteration: example of traditional Chinese medicines

A systematic review of all reports on adulteration of TCMs with synthetic drugs identified 17 case reports, two case series and four analytical investigations of adulteration (Ernst 2002a). The list of adulterants thus generated is impressive and contains drugs associated with serious adverse effects (Table 5).

A recent case of TCM adulteration relates to a 56-year-old man from Indonesia (Huang et al 1997). While visiting Australia, he was admitted to hospital in a confused state which turned out to be due to hypoglycaemia. He insisted that his Type II diabetes was controlled by diet only. However, despite dextrose infusions, his glucose levels would not normalize. Finally, it was discovered that he also took 'Zhen Qi', a TCM bought in Malaysia. It was analysed and shown to contain glibenclamide.

Examples of serious adverse effects caused by such adulterations include agranulocytosis, Cushing syndrome, coma, over-anticoagulation, gastro-intestinal bleeding, arrhythmias, and various skin lesions. The prevalence of adulteration of TCMs cannot be defined at present but one recent report from Taiwan suggests that 24% of all samples were contaminated with at least one conventional pharmacological

TABLE 5   Examples of adulterants found in traditional Chinese medicines

Adulterant	
Aminopyrine	Hydrochlorothiazide
Clobetasol	Hydrocortisone
Dexamethasone	Indomethacin
Dexamethasone acetate	Mefenamic acid
Diazepam	Methyl-salicylate
Diclofenac	Phenacetin
Fluocinolone acetonide	Phenylbutazone
Fluocortolone	Phenytoin
Glibenclamide	Prednisolone

compound (Goudie & Kaye 2001). Adulteration of TCM with synthetic drugs is thus a potentially serious problem that needs to be addressed by adequate regulatory measures.

Conclusion

Herbal medicines are popular but under-researched. Those listed in Table 1 are of proven efficacy and (with two exceptions) their benefits seem to outweigh their risks. For most other HMs, the evidence is too incomplete to even attempt risk-benefit analyses. In order to advise consumers responsibly, we need to fill the current gaps in our knowledge through rigorous research.

References

Bove GM 1998 Acute neuropathy after exposure to sun in a patient treated with St John's Wort. Lancet 352:1121–1122

Canter PH, Ernst E 2003 Multiple n = 1 trials in the identification of responders and non-responders to the cognitive effects of Ginkgo biloba. Int J Clin Pharmacol Ther 41: 354–357

Chan H, Yeh Y-Y, Billmeier GJ, Evans WE 2001 Lead poisoning from ingestion of Chinese herbal medicine. Clin Toxicol 10:273–281

Chan TYK, Critchley JAJH 1996 Usage and adverse effects of Chinese herbal medicines. Hum Exp Toxicol 15:5–12

Cheng TJ, Wong RH, Lin YP, Hwang YH, Horng JJ, Wang JD 1998 Chinese herbal medicine, sibship, and blood lead in children. Occup Environ Med 55:573–576

Chi YW, Chen SL, Yang ML, Hwang RC, Chu ML 1992 Survey of heavy metals in traditional Chinese medical preparations. Chin Med J 50:400–405

Chu JF, Liou SH, Wu TN, Ko KN, Chang PY 1998 Risk factors for high blood lead levels among the general population in Taiwan. Eur J Epidemiol 14:775–781

D'Alauro F, Lin-Fu JS, Ecker TJ 1984 Toxic metal contamination of folk remedy. JAMA 252:3127

Dürr D 2000 St John's wort induces intestinal P-glycoprotein/MDR1 and intestinal and hepatic CYP3A4. Clin Pharmacol Ther 68:598–604

Egger M, Smith GD, Altman DG 2001 Systematic reviews in health care:meta-analysis in context. 2nd edn London: BMJ Books

Eisenberg D, David RB, Ettner SL et al 1998 Trends in alternative medicine use in the United States; 1990–1997. JAMA 280:1569–1575

Ernst E 1999 Second thoughts about safety of St. John's wort. Lancet 345:2014–2016

Ernst E 2000 Risks associated with complementary therapies. In: Dukes MNG, Aronson JK (eds) Meyler's Side Effects of Drugs. 14th ed. Amsterdam: Elsevier p 1649–1681

Ernst E 2002a Adulteration of Chinese herbal medicines with synthetic drugs: a systematic review. J Intern Med 252:107–113

Ernst E 2002b St John's wort supplements endanger the success of organ transplantation. Arch Surg 137:316–319

Ernst E 2004a Randomised clinical trials: unusual designs. Perfusion 17:416–421

Ernst E 2004b Risks of herbal medicinal products. Pharmacoepidemiol Drug Saf 13:767–771

Ernst E, Pittler MH 1997 Alternative therapy bias. Nature 385:480

Ernst E, White AR 2000 The BBC survey of complementary medicine use in the UK. Complement Ther Med 8:32–36

Ernst E, Thompson Coon J 2001 Heavy metals in traditional Chinese medicines: a systematic review. Clin Pharmacol Ther 70:497–504

Ernst E, Pittler MH, Wider B, Boddy K 2006 The desk top guide to complementary and alternative medicine. 2nd edn Edinburgh: Mosby/Elsevier

Escher M, Desmeules J 2001 Hepatitis associated with Kava, a herbal remedy for anxiety. Br Med J 322:139

Espinoza EO, Mann MJ, Bleasdell B 1995 Arsenic and mercury in traditional Chinese herbal balls. New Engl J Med 333:803–804

Giovannini P, Schmidt K, Canter PH, Ernst E 2004 Research into complementary and alternative medicine across Europe and the United States. Forsch Komplementärmed Klass Naturheilkd 11:224–230

Goudie AM, Kaye JM 2001 Contaminated medication precipitating hypoglycaemia. Med J Aust 175:257

Huang WF, Wen KC, Hsiao ML 1997 Adulteration by synthetic therapeutic substances of traditional chinese medicines in Taiwan. J Clin Pharmacol 37:334–350

Ioannidis JPA, Evans SJW, Gøtzsche PC et al 2004 Better reporting of harms in randomized trials: an extension of the CONSORT statement. Ann Intern Med 141:781–788

Johne A, Brockmöller J, Bauer S, Maurer A, Langheinrich M, Roots I 1999 Pharmacokinetic interaction of digoxin with an herbal extract from St John's wort (Hypericum perforatum). Clin Pharmacol Ther 66:338–345

Khawaja IS, Marotta RF, Lippermann S 1999 Herbal medicines as a factor in delirium. Psychiatr Serv 50:969–970

Klassen TP, Pham B, Lawson ML, Moher D 2005 For randomized controlled trials, the quality of reports of complementary and alternative medicine was as good as reports of conventional medicine. J Clin Epidemiol 58:763–768

Koh HL,Woo SO 2000 Chinese proprietary medicines in Singapore. Regulatory control of toxic heavy metals and undeclared drugs. Drug Saf 23:351–362

Lal S, Iskandar H 2000 St John's wort and schizophrenia. Can Med Assoc J 163:262–263

Lawson ML, Pham B, Klassen TP, Moher D 2005 Systematic reviews involving complementary and alternative medicine interventions had higher quality of reporting than conventional medicine reviews. J Clin Epidemiol. 2005 58:777–784

Levitt C, Godes J, Eberhardt M, Ing R, Simpson JM 1984 Sources of lead poisoning. JAMA 252:3127–3128

Li AM, Chan MHM, Leung TF, Cheung RCK, Lam CWK, Fok TF 2000 Mercury intoxication presenting with tics. Arch Dis Child 83:174–175

Lightfoote J, Blair J, Cohen JR 1977 Lead intoxication in an adult caused by Chinese herbal medication. JAMA 238:1539

Markowitz SB, Nunez CM, Klitzman S et al 1994 Lead poisoning due to Hai Ge Fen. The porphyrin content of individual erythrocytes. JAMA 271:932–934

Mathews JD, Riley MD, Fejo L et al 1998 Effects of the heavy usage of kava on physical health: summary of a pilot survey in an Aboriginal community. Med J Aust 148:548–555

Melchart D, Wagner H, Hager S, Saller R, Ernst E 2001 Quality assurance and evaluation of Chinese medicinal drugs in a hospital of traditional Chinese medicine in Germany: a five-year report. Altern Ther 7:S24

Myerson GE, Myerson AS, Miller SB, Goldman JA, Wilson CH 1982 Chinese herbal medicine mail order syndrome. Arthr Rheum 25 (Suppl 4):S87, abstract No. A114

Nemeroff CB, DeVane CL, Pollock BG 1996 Newer antidepressants and the cytochrome P450 system. Am J Psychiatry 153:311–320

Norton SA, Ruze P 1994 Kava dermopathy. J Am Acad Dermatol 31:89–97

Parker V, Wong AHC, Boon HS, Seeman SV 2001 Adverse reactions to St John's wort. Can J Psychiatry 46:77–9

Piscitelli SC, Burstein AH, Chaitt D, Alfaro RM, Falloon J 2000 Indinavir concentrations and St John's wort. Lancet 2000;355:547–548

Pittler MH, Ernst E 2000 Efficacy of kava extract for treating anxiety: systematic review and meta-analysis. J Clin Psychopharmacol 20:84–89

Roby CA, Anderson GD, Kantor E, Dryer DA, Burstein AH 2000 St John's wort: effect on CYP3A4 activity. Clin Pharmacol Ther 67:451–457

Roots I, Johne A, Schmider J et al 2000 Interaction of a herbal extract from St John's wort with amitriptyline and its metabolites. Clin Pharmacol Ther 67:159

Schaumburg HH, Berger A 1992 Alopecia and sensory polyneuropathy from thallium in a Chinese herbal medication. JAMA 268:3430–3431

Schulz V, Hänsel R, Tyler VE 2001 Rational phytotherapy. A physician's guide to herbal medicine. 4th ed. Springer-Verlag; Berlin

Spillane PK, Fisher DA, Currie BJ 1997 Neurological manifestations of kava intoxication. Med J Aust 167:172–173

Stevinson C, Ernst E 1999 Safety of Hypericum in patients with depression: a comparison with conventional antidepressants. CNS Drugs 11:125–132

Stevinson C, Pittler MH, Ernst E 2000 Garlic for treating hypercholesterolemia. A meta-analysis of randomized clinical trials. Ann Intern Med 133:420–429

Stevinson C, Huntley A, Ernst E 2002 A systematic review of the safety of kava extract in the treatment of anxiety. Drug Saf 25:251–261

Stevinson C, Ernst E 2004 Can St John's wort trigger psychoses? Int J Clin Pharmacol Ther 42:473–480

Stoller A 2000 Leberschädigungen unter Kava-Extrakten. Schweiz Ärzteztg 24:1335–1336

Strahl S 1998 Nekrotisierende Hepatitis nach Einnahme pflanzlicher Heilmittel. Dtsch Med Wochenschr 123:1410–1414

Tang JL, Zhan SY, Ernst E 1999 Review of randomised controlled trials of traditional Chinese medicine. BMJ 319:160–161

Tay CH, Seah CS 1975 Arsenic poisoning from anti-asthmatic herbal preparations. Med J Aust 2:424–428

Wu MS, Hong JJ, Lin JL, Yang CW, Chien HC 1996a Multiple tubular dysfunction induced by mixed Chinese herbal medicine containing cadmium. Nephrol Dial Transplant 11:867–870

Wu T-N, Yang K-C, Wang C-M et al 1996b Lead poisoning caused by contaminated Cordyceps; a Chinese herbal medicine: two case reports. Sci Total Environ 182:193–195

Yu ECL, Yeung CY 1987 Lead encephalopathy due to herbal medicine. Chinese Med J 100:915–917

Yue Q-Y, Bergquist C, Gerden B 2000 Safety of St John's wort (Hypericum perforatum). Lancet 355:576–577

DISCUSSION

Azzi: What is the effect of St John's wort on P450? There is a family of P450s, with 57 members in the human genome.

Ernst: It affects the major one, Cyp3A4. This is the one which metabolises most of the prescription drugs.

Azzi: In your adverse effects, you didn't mention that patients may use an herbal medicine rather than a traditional one. In cancer, for example, this is one of the major adverse effects: ignoring traditional medicine.

Ernst: You are correct. In a previous version of this paper I had a separate section on what I call 'indirect safety' aspects. There are plenty of alternative cancer cures. If a cancer patient goes to the internet, they will find 41 000 000 websites advocating complementary therapies. Many of these websites persuade the cancer patient to give up conventional treatment. Under such circumstances even an inherently safe treatment becomes life threatening. There are other indirect safety issues. I could have shown a survey we did with Dr Foster, and organization checking out the track records of doctors and hospitals. We checked out complementary practitioners, and found that the vast majority—even those who are regulated in the UK, such as chiropractors and osteopaths—do not adhere to even the most basic clinical guidelines, such as keeping records and informing the GP. This has previously led me to write that the complementary medicine may be safe (in some cases, at least), but the complementary practitioner certainly isn't.

Katan: Regulators of conventional drugs are very strict with manufacturers of drugs. They don't just go by published papers, but they want more information, and if the drug company falsifies things or doesn't play by the rules, they do so at a serious risk. But there is no such oversight of the manufacturers of alternative therapies. Is there any possibility that data are being falsified, and that there is fraud?

Ernst: Fraud is something associated with human nature. Because research is done by humans, there is no doubt there will be fraudulent data. Part of the issue is what you call 'fraud'. You can do data massaging, or data dredging, or you can call your secondary or tertiary endpoint the primary endpoint simply because this came out positive. There are so many ways of being dishonest. I review a lot of papers for various journals, and perhaps three or four times I have asked to see the raw data. I am ashamed to say that in all cases the journals failed to follow this up.

Russell: How should we move the science forward in the field of herbals and botanicals? One thing that makes doctors in the USA uncomfortable with herbal products is that they don't know what the mechanism of action is. How important is this?

Ernst: For some, we do suspect mechanisms. For instance, for St John's wort we think it is acting as a selective serotonin reuptake inhibitor (SSRI). There is some evidence there. Even for St John's wort there is a degree of uncertainty. There is more than one family of compounds believed to be pharmacologically active. If we isolated just one compound and took it out the chance is that it would not have the same effect. Sometimes this works: it has worked for aspirin, which comes from willow bark. In most cases there is a postulated synergy: herbalists frequently believe that these various compounds work with different mechanisms of action at the same time.

Russell: Couldn't that be researched?

Ernst: Yes, but what a task that is. It is an insurmountable task considering the paucity of research funds.

Scott: On your benefit slide you didn't have the placebo effect. Apart from cases where people have been diverted away from proper treatment, isn't it a fact that a lot of people derive benefit from the placebo effect?

Ernst: That's a good point. It's a complex issue. I'm sure that the placebo effect for some of these treatments is enormous. People pay out of their own pockets, and the more you pay the more it is worth. People have very empathetic encounters with therapists, and so forth. As an ex-clinician I know that the placebo effect can be powerful, and there's nothing wrong with profiting from this. But you don't need an ineffective treatment to benefit from a placebo effect: effective treatments also have a placebo effect. They come as a bonus with everything, so you might as well prescribe something effective for your patient.

Rayman: How do you deal with the quality differences in preparations? You must be systematically reviewing stuff that is different from one trial to the next.

Ernst: I tried to address this. We have conducted reviews where we looked at different extracts, such as St John's wort or garlic. It didn't make all that much different. You need enough data to do this. Usually we have seven or eight trials of three different extracts, and then you can't subgroup according to extract because the subgroups become too small. Suboptimal extracts can only produce a false negative overall result. Also, who would do a clinical trial with an extract that is worthless? By and large people try to use something that is considered to be the optimum extract.

Yetley: I noticed on your list of products for which you have conducted evidence-based reviews that you included ephedra and soy. These have been the subject of recent reviews in the USA. How similar are your conclusions to those from the US reviews?

Ernst: This was quoting from our book, and they are identical. For ephedra we took Shekelle's meta-analysis. We didn't re-do the meta-analyses where they were available. It was systematic in the sense in that we included systematic reviews where they were available, and we ourselves had published a lot. When no review was available, we reviewed it for that particular book chapter.

Taylor: Can you comment on research priorities for characterizing these substances, in terms of their chemical characteristics?

Ernst: That's a difficult question. I would rely on the experts; I can't be an expert across the board.

Taylor: Isn't part of the challenge that so many of these alternative remedies are not just a single ingredient?

Ernst: The problem is huge, because in many cases we don't know what the active ingredient(s) is. This has implications for standardization. Producers do standardize, but sometimes this is for what is standardizable rather than for what is active. They standardize for marker substances. It quickly gets so complicated, so that for each extract you would have to convene a panel of different experts on that particular issue. Another related issue is that when we review adverse effects we rely on case reports. Having done this for a while I know that case reports in the literature are often terrible. Even the best ones don't allow you to conclude causality or probability of causality. Currently we are trying to get a checklist or guideline for publishing reasonably good case reports.

Russell: Are the standards of identity for these herbal medicines based on one marker substance, or a series of marker substances? What are the criteria for standards of identity that should be used before starting a clinical trial?

Ernst: To the best of my knowledge there is no standard. Across the board, for each extract they have, by convenience, usually found one. Modern technology would allow us to go much further than this but it hasn't been implemented.

Klein: There was a comment earlier about mechanistic studies. The National Center for Complementary and Alternative Medicine has been supporting studies for seven years or so now. Over 50% of our studies are clinical. They are clinical because people are using these products and they want to know whether they are safe. We are trying to shift this proportion such that more studies are mechanistic. But as has been mentioned there are some difficulties getting down to the level of mechanism. Another comment was made about the placebo effect. We actually solicited studies to understand the placebo effect. If we can tease out what the mechanism of the placebo is, this could benefit conventional medical practice. Regarding product quality, over the last year we have implemented a new policy, whereby before supporting any research the products had to go through an additional evaluation besides our peer review process: they are now subjected to the scrutiny of a small group of botanists, natural product chemists and microbiologists. We have published the criteria we want the investigators to address when they submit product information to us. For example, for plant products we want the taxonomic name, the source of the original plant material, where it was grown, how it was harvested, the time of year that it was harvested and information on extraction method, among others. The investigators have to provide a plan for reserving product and analysing it. For many of the products we don't know what the active ingredient is. We hope that for some time after publication of results the investigators will still have product for further analysis should new information arise. In addition, we ask that investigators proposing clinical studies approach the FDA regarding the need for an IND (investigational new drug) application, and if the FDA insist on this then they have to proceed with this. It is another measure of quality. Then that study is subjected to protocol issues in addition to product quality issues. We have an IND for cranberry juice cocktail, so it is not just dietary supplements but functional foods as well.

Ernst: I think these criteria are quite reasonable. Any responsible manufacturer should have this information. Should you as a primary investigator confirm it independently? I think so, yes. Sadly, all of this paperwork and analytical work increases the cost of a study tremendously.

Halliwell: One comment that I thought was quite profound is that the herbal preparations that get put into studies are the ones that people think are likely to work. It also struck me that the only reason people adulterate preparations with pharmaceuticals is because they know 'in their hearts' that they don't work. To me, this means that 80% of these herbal remedies don't work except possibly by the placebo effect. As a manufacturer you won't put your preparation through a trial if you know it doesn't work. Is that a fair comment or am I being over-critical?

Ernst: I wouldn't say 80%; I would say 'a proportion'.

Azzi: Your analysis of risk–benefit also included documentation of the producer. If you don't include that information, are your conclusions different? For example, for vitamin E, different producers make the following claims on the internet: it protects against cancer, atherosclerosis, neurodegenerative diseases, erectile dysfunction and so on. We know that these claims are false.

Ernst: What I tried to express is that depending on a subject, if you study a traditional Chinese herbal medicine and don't include the Chinese literature, you are not being systematic. We have included the Chinese literature a few times, analysing over 2000 so-called RCTs. We found that the vast majority (>90%) are not worth the paper they are printed on. For instance, we found RCTs without a control group.

Azzi: If you don't include the information which you judge to be useless, does this change the result?

Ernst: You don't include useless information into a meta-analysis. If you want to consider the Chinese literature you have to discard a lot of the studies.

Katan: You told us that herbal remedies are by and large safe. Could you expand on this? Is this from case reports or clinical trials?

Ernst: The basis for this statement is the notion that if something has survived 3000 years it doesn't kill too many people. I used cautious terms because this doesn't prove anything.

Katan: If a herbal remedy caused common chronic diseases like heart disease or colon cancer you would never detect it. A case in point is the cholesterol-raising factor from coffee beans. Scandinavians have been drinking their coffee including this factor for about 150 years. We know now that this raised cholesterol has caused heart disease in many people (Tverdal et al 1990), but it didn't become obvious until scientists specifically investigated Scandinavian coffee, first by epidemiology and then by clinical trials (Weusten-Van der Wouw et al 1994). I think a history of safe use will fail to pick up food components which cause common chronic diseases Even cigarettes, which raise the risk of lung cancer 15-fold, weren't identified as a problem until there was systematic research.

Ernst: You are absolutely right for challenging me on this. I would attack myself if I were sitting where you are. But there is a bit of rudimentary epidemiology. There is the Uppsala WHO centre that collects adverse effects from more than 50 countries. They have a mountain of data and they look for signals. I am one of their herbal reviewers. We are trying to pick out a signal. It is by no means foolproof.

Rayman: I have recently looked at the literature on arthritis. In included herbal medicines. It was refreshing and helpful to have the kinds of reviews that Edzard Ernst has produced. There are too few people doing this sort of thing.

Aggett: I think there's going to be an emerging confluence between your approaches and potential effects with natural or regularly consumed foodstuffs such as brassicas. We mustn't set the herbal remedies totally aside, because there is a continuum. One issue that's a current hot topic is the issue of natural toxicants, and how these may respond to environmental influences, in particular, radiation and global warming.

Ernst: I hope there is also a confluence of research funds.

References

Tverdal A, Stensvold I, Solvoll K, Foss OP, Lund-Larsen P, Bjartveit K 1990 Coffee consumption and death from coronary heart disease in middle aged Norwegian men and women. Br Med J 300:566–569
Weusten-Van der Wouw MP, Katan MB, Viani R et al 1994 Identity of the cholesterol-raising factor from boiled coffee and its effects on liver function enzymes. J Lipid Res 35:721–733

Standardization and evaluation of botanical mixtures: lessons from a traditional Chinese herb, *Epimedium*, with oestrogenic properties[1]

E. L. Yong, S. P. Wong, P. Shen, Y. H. Gong, J. Li and Y. Hong*

*Department of Obstetrics & Gynaecology, National University Hospital, Yong Loo Lin School of Medicine, National University of Singapore, Lower Kent Ridge Road, Republic of Singapore 119074, and *Temasek Life Sciences Laboratory, National University of Singapore, Singapore*

Abstract. Botanical extracts differ from conventional supplements in that they are complicated mixtures of many bioactive compounds. Here we describe our experience with a traditional Chinese medicinal plant *Epimedium* sp. to illustrate the scientific challenges of firstly, obtaining a standardized product from a complex mixture and secondly, evaluating that product for preclinical and clinical efficacy. In contrast, to its colloquial name '*Horny goat weed*' and internet advertisements as a herbal 'Viagra' for men, extracts of *Epimedium* are strongly oestrogenic due to the presence of novel potent phytoestrogens of the prenyl-flavone family. Since *Epimedium* is not cultivated, it was necessary to source for taxonomically identified samples and to authenticate their species by phylogenetic, chemical and bioresponse profiling. The feasibility of using a panel of oestrogen-responsive cell-based bioassays to measure summated oestrogenic effects at close time points for pharmacokinetic/pharmacodynamic (PK/PD) modelling was evaluated. We document proportionate oestrogenic responses in sera of animals fed oestrogenic drugs and botanical extracts, indicating that these target molecule responsive cell-based bioassays may have utility to capture the global effects of the myriad bioactive compounds in botanical extracts, informing the design of rigorous clinical trials for safety and efficacy.

2007 Dietary supplements and health. Wiley, Chichester (Novartis Foundation Symposium 282) p 173–191

Botanical supplements originating from folk medicines are extremely popular. Traditional Chinese medicines (TCMs) have the longest and most sophisticated documentation among folk medicines, and its herbal formulations are used regularly by a majority of people in the Far East and significant numbers globally. A survey among preoperative patients in Hong Kong indicates that 90% of patients used Chinese herbs on a regular

---

[1] Some of the data discussed in this chapter have been published in Shen P, Guo BL, Gong Y, Hong Y, Yong EL 2007 Taxonomic, genetic, chemical and estrogenic characteristics of Epimedium species. Phytochemistry 68:1448–58, and Wong SP, Li J, Shen P, Gong Y, Yap SP, Yong EL 2007 Ultrasensitive cell-based bioassay for the measurement of global estrogenic activity of flavonoid mixtures revealing additive, restrictive and enhanced actions in binary and higher order combinations. Assay Drug Dev Technol DOI:10.1098/adt.2007.056.

basis and over 40% had consulted a TCM practitioner within the last 12 months (Critch-ley et al 2005). The State Pharmacopoeia Commission of PR China (2000) lists over 500 individual plants in its *materia medica*. *Epimedium* sp. (*Berberidacea*) is one of the most popular 'yang' tonic herbs and its use dates to the ancient text, Shen Nong Ben Cao Jing (*c.* 200 BC–100 AD). The herb is listed as having action to 'reinforce the kidney *yang*, strengthen tendons and bones, and relieve rheumatic conditions' and is indicated 'for impotence, seminal emission, weakness of the limbs, rheumatoid arthralgia with numb-ness and muscle contracture, and climacteric hypertension'. *Epimedium* sp. consists of the dried leaves of several different species including *E. sagitatum* (Sieb. Et Zucc), *E. koreanum* Nakai, *E. pubescens* Maxim, *E. wushanenese* T.S. Ying and *E. brevicorum* Maxim (The State Pharmacopoeiea Commission of PR China 2000). The herb is an evergreen shrub indig-enous to shady mountain areas in temperate to subtropical Asia. Dried leaves of *Epime-dium* can be boiled with water to make a decoction, or it can be macerated in wine for oral consumption (Li *c.* 1500). It is also used as part of complicated formulations with other herbs such as rehmania root, curculigo rhizome, dogwood fruit and wolfberry fruit.

In vitro studies indicate that *Epimedium* extracts and its chemical constituents can exert pharmacological effects on cells of the immune, cardiovascular, neuronal, res-piratory, genital/urinary, hepatic, endocrine, and skeletal systems (reviewed in Yap & Yong 2005). A major focus for use of *Epimedium* herb is in the treatment of osteoporosis (Table 1). *Epimedium* was reported to enhance the osteogenic differ-

TABLE 1   Effects of *Epimedium* on bone health

***In vitro* studies**

Source	Assay	Test substance	Conclusions
Liu 1984	Bone marrow cell cultures of 'yang deficiency' animal model caused by hydroxyurea	*E. sagittatum* polysaccharides	*E. sagittatum* polysaccharides increased cell multiplication and DNA synthesis of bone marrow cell cultures
Li et al 2002	Osteoclasts from rabbit bones	*Epimedium*	Induces osteoclast apoptosis by TUNEL staining
Yin et al 2005	Human osteoblasts induced from mesenchymal stem cells	*E. pubescens* Icariin	Enhances proliferation and differentiation of cell cultured human osteoblasts through increasing cell cultures BMP2 mRNA
Meng et al 2005	Osteoblast-like UMR106 cells	*E. brevicornum* compounds (icariin, epimedin B, C)	Icariin promoted proliferation of osteoblast-like UMR106 cells
Chen et al 2005	Primary culture of rat bone marrow stromal cells	Icariin	Increased the alkaline phosphatase activity, osteocalcin secretion and calcium deposition

TABLE 1   (*Continued*)

Animal studies

Source	Test substance	Subjects/Model	Response
Wu et al 1996	water extract of Herba Epimedii	rats	Herba Epimedii can prevent the side effects induced by long-term use of glucocorticoids in rats.
Peng et al 1997	*E. leptorrhizum*	Male Sprague-Dawley rats	*E. leptorrhizum* has no significant effects on endogenous cAMP in alveolar bone of orthodontic tooth in 11 rats.
Yu et al 1999	*E. leptorrhizum* Stearn; oestradiol	Ovariectomized rats	Both the *Epimedium* and oestradiol were able to increase mineral content and promote bone formation.
Ma et al 1999	*Epimedium*	Wistar ovariectomized rats	Trabecular volume in total bone volume (TV/TBV), and osteoid percentage increased significantly after epimedium treatment.
Wang et al 2000	*Epimedium*	Ovariectomized rats	The BMD was significantly higher, but the IL6 mRNA expression level was significantly lower in epimedium group than that in the OVX group.
Jiang et al 2002	*Epimedium* flavonoids	Rat castrated osteoporosis model	Flavonoids increased bone mineral density as measured by DEXA and increase oestradiol and decreased IL6 levels in serum.
Chen et al 2004	*E. sagitttatum* flavonoids	Effects of serum from rats fed flavonoids	Flavonoid fed-serum increase proliferation and differentiation of rat calvarial osteoblasts compared to controls, whereas original flavonoid was inactive.
Zhang et al 2006	*E. brevicornum* Flavonoids	Ovariectomized Wistar rats	Flavonoids prevented bone resorption, stimulated bone formation, and prevented osteoporosis without hyperplastic effect on uterus.

TABLE 2   Chemical contents of *Epimedium* sp.

Family	Compounds
Flavonoids	luteolin, chrysoeriol, quercetin, apigenin, apigenin 7,4″dimethyl ether, kaempferol, thalictoside, brevicornin, epimedokoreanoside 1, yin yang huo A–E
Prenyl-flavone glycosides	Icariin, Baohuoside 1- epimedokoreanoside I, icariside I, icaritin, epimedoside A, epimedins A, B and C, β-anhydroicaritin, tricin, korepimedoside A and B, sagittatosides A–C, sagittatins A and B, diphylloside A and B, baohuosides I–VII and baohuosu, Desmethyicaritin, Baohuoside, Iskarisid II, Iskarisiside A, Phenoxychromones
Sterols	β-sitosterol
Essential oils and fatty acids	linolenic acid, oleic acid, palmitic acid, sterols, tannins
Vitamins and minerals	vitamin E, zinc
Polysaccharides	

entiation of rat primary bone marrow stromal cells (Chen et al 2005), increase osteoblastic proliferation (Meng et al 2005), reduce osteoclastic bone resorption (Yu et al 1999) and to increase mineral content, and prevent osteoporosis in ovariectomized rats (Zhang et al 2006). These data make plausible the hypothesis that *Epimedium* decoctions may prove efficacious in treatment of osteoporosis in humans.

Prenyl-flavones in *Epimedium* are potent phyto-oestrogens

In view of its purported effects on bone health, our laboratory investigated whether extracts of *Epimedium* and its constituent compounds can activate the oestrogen receptors (ERα and ERβ), part of a 48-member family of transcription factors, which regulate bone and skeletal health in women. An ethanol extract of *Epimedium* was screened for oestrogenic activity with cell-based oestrogen-sensitive reporter gene assays. This herb, and not 11 others with purported 'ying–yang effects', induced strong and specific oestrogenic effects (Yap et al 2005, De Naeyer et al 2005). Despite their oestrogenicity, *Epimedium* extracts have dose-dependent stimulatory (Wang & Lou 2004) and inhibitory (Yap et al 2005) effects on the proliferation of oestrogen-responsive MCF-7 breast cancer cells. We performed purification steps using bioassay-guided column chromatography and repeated preparative HPLC separations to define the bioactive entities responsible for oestrogenic effects of *Epimedium*. A novel prenyl-flavone, breviflavone B, was isolated (Table 2) (Yap et al 2005). Prenyl-flavones are a recently described class of flavonoids with potent oestrogenic properties. An example is 8-prenyl-naringenin, a strong phyto-oestrogen first isolated from hops and beer, which shows binding activity to both ERα and ERβ (Kitaoka et al 1998, Milligan et al 2000, Schaefer et al 2003). Intriguingly, members of this class of compounds exhibit partial anti-oestrogenic activity in the female rat uterus and MCF-7 human breast cancer cells (Pedro et al 2006) and may be specific inhibitors of the breast cancer resistant protein ABCG2 (Ahmed-Belkacem et al 2005). Our compound, breviflavone B displayed strong oestrogenic activity and was also able to inhibit the growth of ER-responsive breast cancer cells at high doses. Interestingly, this inhibitory activity was associated with reduction of ERα protein. In comparison, presence of the selective oestrogen receptor modulator (SERM) tamoxifen increased ERα protein content. Reduction in ERα protein was reversible by the proteasome inhibitor, lactacystin, indicating protein degradation through ubiquitin proteasome pathways (Yap et al 2005). In this respect, the action of breviflavone B resembles the action of the pure anti-oestrogen, ICI 182,780, which binds ERα with high affinity and induces a rapid proteasome-dependent degradation of the receptor (Long & Nephew 2006).

The oestrogenic properties of *Epimedium* coupled with these anti-proliferative effects on breast cancer cells suggest its possible utility for oestrogen replacement therapy, but without the adverse effects on breast health associated with current oestrogen–progesterone formulations. We were keen to examine whether *Epimedium* herb prepared in the traditional manner can exert any oestrogen effects in the human under the standardized conditions of a randomized controlled trial. In our efforts to organize this first clinical trial, we encountered several challenges that we believe many researchers investigating traditional Chinese herbal medicines will face.

Standardized herbal raw material

The prerequisite for a scientific study is the procurement of standardized drugs with reproducible pharmacological properties. A major problem with regards to *Epimedium* is the absence of agreement on the precise number of *Epimedium* species. Chinese taxonomists and geneticists have variably reported numbers ranging from 20 to 50 species (Sun et al 2005). Traditional herbalists do not differentiate among *Epimedium* species, but rather use the genus together as *Herba Epimedii* (Yang 1985). Raw materials from even reputable herbal wholesalers in Singapore are not taxonomically identified with regards to species. Inspection of these materials revealed leaves of different shapes, morphology and texture suggesting that they consist of a mixture of species. These species differ significantly in concentrations of major and minor constituents (Wu et al 2003). Another problem is that *Epimedium,* like the majority of TCM herbs, is not cultivated but is collected from the wild, which increases the dangers of wrong species identification, genetic diversity, and possible differences in levels of bioactive compounds due to soil and climate (Guo & Xiao 2003). Other factors that may cause variations include differences in processing, packaging and storage of raw materials.

Taxonomic identification of herbal raw materials

To obtain authenticated herbs, it was necessary to engage an herbal taxonomist. Qualified taxonomists tend to specialize in one (or a very narrow) group of plants and are an increasing rare group of specialists. Since there are over 500 commonly used herbs in the Chinese Pharmacopoeia, it might not be a straightforward task to locate a collaborator with the necessary expertise. Fortunately Dr Guo Bao Ling, of the Institute for Medicinal Plant Development, Chinese Academy of Medical Sciences, Beijing agreed to assist us (Guo & Xiao 1996, 2003). Specimens of *Epimedium* need to be collected at springtime, when the corolla characteristics of flower petals can be used to aid identification (Fig. 1). Even so, definitive

FIG. 1.   A voucher specimen of *Epimedium sagittatum* (Sied et Zucc) Maxim. Taxonomic identification is helped by the presence of distinctive flowers. Other features include the three leaflets on each stalk and presence of sharp pointed leaves. Specimen was supplied by Dr Liu Ruoyong, Henan Chinese Traditional Medicine College.

identification is rather problematic and the proliferation of species may reflect the uncertainty as to which characteristics are most specific. Nevertheless, there appears to be general agreement that plants growing in the north of China and Korea are *E. Koreanum*, as the other species do not thrive in such cold conditions (Guo & Xiao 2003). As discussed later, DNA profiling confirmed that *E. koreanum* was indeed distinct from the other species (Sun et al 2005). Specimens collected in Sichuan province of China were thought to belong to two closely related species, *E. pubescens* or *E. brevicornum*. Their close relationship begs the question whether they are different species or may be more appropriately classified as variants of one species. Plants originating from the eastern and south of China are the most problematic as several species may coexist in the same area. Because of these difficulties, we are embarking on a systematic collection of *Epimedium* sp. from different regions of China and subjecting them to DNA and chemical profiling.

DNA profiling

Genetic differences, or differences in genomic DNA, are regarded as definitive means of botanical identification compared to morphology. These genetic tools include DNA fingerprinting using multi-loci probes, random amplified polymorphic DNA (RAPD), restriction fragment length polymorphism (RFLP), amplified fragment length polymorphism (AFLP), microsatellite marker technology and sequencing of poorly conserved regions of nuclear ribosomal DNA. Using the last technique, we sequenced an 898 bp fragment from an internal-transcribed spacer between regions encoding 18S, 5.8S ribosomes in an *Epimedium* species. Genetic differences between the samples from difference sources were observed in 9 loci (Table 3). However, differences in oestrogenic activity were observed even in the absence of genetic differences (comparing *E. brevicornum* samples from Henan and Ruoyang with Eu Yan Sang). To obtain more comprehensive genetic data, we used AFLP analysis to examine differences across the whole plant genome. DNA was extracted from 37 *Epimedium* samples and fluorescent AFLP analysis was performed. Restriction fragments were used to construct a phylogenetic tree and preliminary data are shown in Fig. 2. Most species are genetically distinct with samples of the same species grouped together. *E. koreanum* was found most distinct from other species. Intra-species variation was also very clear for some species. The results indicate that most genetic variation was due to inter-species variation. Intra-species variation differed among species and warrants further investigation.

Standardization and quality assurance is difficult for wild-grown species like *Epimedium*. On the other hand, genetic homogeneity cannot be taken for granted even in farm-grown herbs. There is a recent report on genetic and chemical variation among *Panax notoginseng* samples from one single farm (Hong et al 2005).

TABLE 3 Genetic differences in *Epimedium brevicornum* samples from different sources

Epimedium samples	Nucleotide positions									Relative oestrogenic bioactivity
	112	279	299	306	493	597	607	662	680	
EB Germany	C	G	G	A	A	A	G	A	T	+++
EB Henan	T	G	G	G	A	G	G	G	T	++
EB Ruoyang	T	G	G	G	A	G	G	G	T	++
EB Eu Yan Sang	T	G	G	G	A	G	G	G	T	++++
EB (GenBank)	T	A	G	A	A	G	C	G	C	NA

DNA was extracted from leaves of herb samples and sequencing of a 898 bp fragment of an internal-transcribed spacer region from a gene encoding ribosomal proteins was performed. Genetic differences were observed in 9 loci and these were compared to oestrogenic bioactivity as measured with oestrogen-driven reporter gene bioassays. Reference sequence was obtained from GenBank. NA, not available.

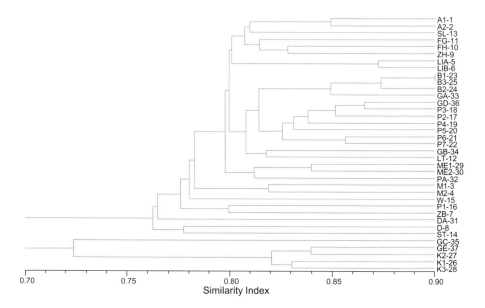

FIG. 2.   DNA profile of *Epimedium* species. Samples of *Epimedium* were collected from locations all over China by Dr Guo Bao Ling, Institute for Medicinal Plant Development, Beijing. Amplified fragment length polymorphism (AFLP) analysis was performed and differences in restriction sites were used to construct this phylogenetic tree. A total number of 194 restriction fragments were used to calculate the similarity index (1.00 being exactly identical). The use of more restriction sites may result in phylogenetic trees that may be more closely related to pharmacological activity.

P. notoginseng has been cultivated for over 1000 years in Wenshan country, PR China. *P. notoginseng* roots were obtained from a single farm and processed in a standardized manner for AFLP analysis. 159 discrete polymorphic bands were obtained and used for phylo-genetic analysis. Genetic variation was clearly observed even with plants from a single farm. HPLC and HPTLC analysis also revealed variation in saponin constituents, their main active compounds. There is, however, no strict correlation between genetic variation and chemical component variation, indicating the necessity for chemical and bioresponse profiling to supplement DNA analysis.

Chemical profiling

Major compounds in *Epimedium* are the flavonol glycosides, icariin, Epimedins A, B and C and are frequently used as biomarkers (Yap & Yong 2005). Flavonol glycosides in the leaves of *E. grandiflorum, E. cremeum* and *E. sempervirens* vary with seasons (Chen et al 1996). Total content of glycosides was highest in quantity at the flowering time, and as the leaves mature it became less variable. The ideal period for the harvest of *Epimedium* leaves is two or three months after the flowering time. In *E. koreanum*, the highest content of flavonoids was found in the flowering period (Mizuno et al 1989). By means of RP-HPLC, nine major flavonoids in different parts of five *Epimedium* plants were analysed. Total content of nine flavonoids in the four species were found to be highest in the rhizome and roots, followed by leaves and stems (Guo & Xiao 1996). The composition of the main constituents and relative contents in the five species were similar in leaves and stems, but different from the rhizome and roots. These differences may affect their pharmaceutical properties indicating the importance of specifying harvesting time as well as which part of the plant will be used in any clinical study. However icariin glycosides were not oestrogenic in our assays. On the other hand, the bioactive prenyl-flavone aglycones, 3'-prenylapigenin and breviflavone B, were present in less than 0.1% of the ethanol extract (Yap et al 2005). Besides breviflavone B, phytoestrogens present in small quantities include luteolin, kaempferol and apigenin (Wu et al 2003). Thus, measurement of the major compounds alone would be misleading for standardization of oestrogenic activity, a point we shall return to in our discussion on biological response profiling.

Relationship between phylogenetic and chemical profiling

The *P. notoginseng* study (Hong et al 2005) indicates that even plants from a single farm, where defined parts of the plant were harvested and processed under standardized conditions, did not exhibit complete uniformity of chemical constituents. Using HPTLC and HPLC to analyse the relative abundance of the four major saponins Rg1, Rb1, R1 and Rd, it was found that although genetic diversity had contributed to the variation of saponin contents, there was, however, no strict correlation between abundance of saponins with any particular genetic makeup. Several factors may explain such loose linkage. Firstly, most genetic changes occur at non-coding DNA sequences and these non-coding sequences comprise the bulk of genomes in plants. Secondly, many changes in coding regions are silent and do not lead to amino acid change. Thirdly, only a relatively small number of proteins from the entire proteome are involved in saponin biosynthesis and these may not be affected by the genetic variances detected.

While there have been intensive breeding activities for crops, there has been little attention paid to the breeding of medicinal herbs. One of the reasons for this may be a lack of understanding as to which characteristic is important for its pharmacological properties. Farmers rely on morphologic characteristics for their breeding activities. Since the desired pharmacological properties are readily quantified and if known may not be appreciated by the consumers, efforts were directed to the production of bigger plants with heavier roots which can fetch a higher market price. One possible strategy for breeding *P. notoginseng* may be based on the levels of Ginseng saponins. Since Rd is potentially involved in herb–drug interactions, cultivars with lesser amounts of Rd may be desirable. Cultivars with a higher relative ratio of R1 could be more effective in treating cardiovascular diseases. Furthermore since Rb1 and Rg1 have opposing properties in angiogenesis, more Rg1 would be needed for wound healing while the inhibitory activity of Rb1 will be useful for cancer treatment. Such an approach presupposes a broad understanding of the pharmacological effects of chemical constituents of the herb. Unfortunately, for most herbs, elucidation of mechanisms of action of individual compounds is far from complete. Combinatorial effects of the many bioactive compounds in an herb or herbal formulae add additional layers of complexity.

Bioresponse profiling

The unsatisfactory correlation between phylogenetic and chemical profiling necessitates the design of tools to capture summated effects of the compounds. In our investigation into the biological effects of *Epimedium*, we used a promoter with oestrogen response elements driving a luciferase gene. Such reporter genes when transfected into appropriate cell lines are accurate and sensitive biomarkers of the action of oestrogens. They have near linear responses over several orders of magnitude (Wang et al 2005). These tools allow the possibility of examining different promoters in appropriate cellular contexts. Oestrogen-responsive reporter gene assays have been shown to correlate with whole animal responses such as increases in thickness of uterine lining (Sonneveld et al 2006). Since we have used these bioassays to identify the oestrogenic activity of *Epimedium*, we hypothesize that they may be good tools to complement genetic and chemical profiling in defining herbal raw materials. The use of such tools may even be useful for studies to determine preclinical efficacy. We have examined the oestrogenicity of ethanol extracts of different *Epimedium* species prepared under standardized conditions. There was a reproducible hierarchy of bioactivity with ERα-responsive gene bio-assays. Studies are now in progress to compare ERα and ERβ effects in different promoters, and their interactions with coactivators and corepressor molecules in cells from clinically relevant organs such as the breast, uterus and bone. Such an approach has the potential to reveal whether botanical mixtures have any selective oestrogen receptor modulator (SERM) properties with beneficial effects on bone without adverse effects on breast and endometrial health.

Pre-clinical and clinical evaluation

Clinical evaluation of botanical extracts is complicated by the absence of basic pharmacokinetic/pharmacodynamic (PK/PD) data that is necessary to formulate medicines with appropriate absorption, distribution, metabolism and excretion characteristics. Current methods to determine PK/PD data rely on mass spectro-scopic measurements of serum concentrations of one (or a small number of) bioac-tive compound. A platform to measure the summated effects of numerous bioactive compounds with possible additive, synergistic or antagonistic properties are required, since botanical extracts have hundreds (possible thousands) of bioactive compounds. We have evaluated the feasibility of using a panel of oestrogen-responsive cell-based bioassays to supplement mass spectrometric measurement of bioactive compounds. Male Sprague-Dawley rats were administered the standard oestrogenic pro-drug oestradiol valerate and standardized ethanol *Epimedium* extract and serum collected over 5 hours for measurement of oestrogenic activity. Bioactive small molecules were extracted from serum and tested (1) with ERα and a vitellogenin ERE–luciferase reporter transfected into Hela cells, and (2) effects on oestrogen-responsive breast cancer cell proliferation. Oestradiol valerate induced proportionate oestrogenic responses in the ERα/ERE–luciferase reporter and breast cancer cell bioassays. In contrast, the extract admin-istered orally and injected induced asymmetrical responses in the two bioassays, suggesting the possibility that these bioassays may be useful for determining SERM activities. Oestrogenic effects were distinct in terms of effects on different ERE promoters in different cell lines, providing a mechanistic basis for under-standing SERM action. Our data indicate that the oestrogenic action of some fla-vonoids are additive, whereas partial antagonistic effects can be demonstrated for others. These target molecule responsive cell-based bioassays may have utility to capture the global PK/PD effects in animal and human studies, informing the design of definitive clinical trials for human safety and efficacy.

Safety and standardization

Traditional texts have been through much iteration over the centuries and describe many possible methods of preparing *Epimedium*. The main ways appear to be boiling and soaking in alcohol; the latter method is considered 'to be more effective'. Examination of water compared to alcohol extracts indicated that the dose-response curve of the water extract was shifted to the right with respect to alcohol (EC_{50}: 25 versus 150 µg/ml) (Fig. 3). Extracts with more potent oestrogenic activity (EC_{50}: 1.3 µg/ml) were obtained by further purification of the bioactive prenyl–flavone fraction (Yap et al 2005). However, such purified extracts are not the same as those prepared by traditional methods. Would a highly purified fraction consisting of strongly oestrogenic prenyl-flavones still have the same safety/toxicity profiles as a decoction that adheres to centuries old methods? Guidelines from regulatory agencies are unclear on this point. The US FDA Guidance on Botanical Drugs states that highly purified extracts will be considered on a case-by-case basis (US Department of Health and Human Services 2004). In Germany, highly purified extracts are considered new chemical entities that have to undergo toxicity testing in a manner similar to a compound newly synthesized in the laboratory. Scientific forums such as this symposium have a key role to play in forming regulatory opinion. Potentially valuable botanical medicines that have stood the test of time should be brought to the consumer safely, but without superfluous testing. Nonetheless safety for *Epimedium*, like any other herbal extract, depends on the availability of raw materials that have undergone authentication by

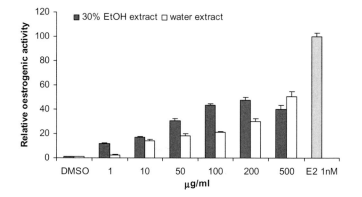

FIG. 3.   Oestrogenic activity of ethanol and water extracts of an *Epimedium* species. Extracts were prepared from leaves of *E. pubescens* by soaking in 30% ethanol for 3 days, or boiling in water for 3 hours. Extracts were dried and reconstituted for testing with a HeLa cell line stably transfected with ERα and a luciferase reporter gene driven by an oestrogen-responsive promoter. Relative estrogenic activity (±SE) were expressed as percentages of a saturating dose of 17β-oestradiol (1 nM).

taxonomical, genetic, and chemical profiling and which are made to good manufacturing practices.

Manufacturing and regulatory agencies

Like other dietary supplements, *Epimedium* is currently sold to consumers in dozens of different preparations. Many of these medicines use extracts made in China, which are then encapsulated in the USA to meet regulatory requirements (personal communication with manufacturers). Under such circumstances, integrity of manufacturing processes and chemical contents cannot be guaranteed. Unfortunately, adulteration can occur, such as documented for PC-SPES, an extract that seems promising for management of advanced prostate cancer (DiPaola et al 1998). PC-SPES caused gynaecomastia, deep vein thrombosis and bleeding diathesis due to adulteration with the synthetic drugs diethylstilbestrol and warfarin (Sovak et al 2002). In Singapore and elsewhere (EMEA Committee for proprietary medicinal products 2001), herbal products can be sold over the counter provided the following minimal standards are met. The products should not have toxic levels of heavy metals and microbial contamination and adulteration with a list of synthetics chemicals are not present. We have tried to engage several manufacturers in Taiwan, mainland China, the USA and Europe to manufacture *Epimedium* extract to above minimal standards for a clinical trial. Most of them were reluctant to embark on quality manufacturing because of the costs and uncertainty of returns on the investment. They point to competition from products already on the market. It would be difficult for a quality *Epimedium* extract to emerge unless, and until, consumers demand this and regulatory authorities make it advantageous for manufacturers to do so.

Summary

Legend has it that *Epimedium* received its colloquial name, 'Yinyanghuo' or 'horny goat weed', when goats grazing on the herb were observed to have excessive copulating behaviours. The herb is considered an aphrodisiac and is used to treat sexual dysfunction in both men and women. In contrast to common belief, we found extracts of *Epimedium* to be strongly oestrogenic due to the presence of potent prenyl-flavones. The lack of standardization and scientific evidence for efficacy is leading to a decline in sales for botanical supplements. Even in China, the preference is for modern pharmaceutical compounds because of the low quality of TCM medicines. However TCM botanicals have centuries of human use, have stood the test of time and are relatively cheap. A growing literature (albeit mostly *in vitro* data), attest to the potential value of compounds present in these extracts. Although having generally lower potencies compared to pharmaceutical compounds, they may have similarly milder adverse effects. Moreover, medicines with lower potencies may be desired for chronic conditions associated with ageing populations such as diabetes, cardiovascular disease and for supplementation of hormone deficiencies. Mixtures are also different from pure compounds. The many different metabolites from a common backbone present in botanical extracts may lead to lower toxicity. Summated effects may mean that lesser amounts of any single one compound are required. If creative solutions can be found to problems of standardization and clinical evaluation of complex botanical mixtures, scientific study of this important resource of natural medicines would be greatly improved.

Acknowledgements

We are grateful to the National University of Singapore Academic Research Fund, and US National Institutes of Health and National Center for Complimentary and Alternative Medicines for partial funding support.

References

Ahmed-Belkacem A, Pozza A, Munoz-Martinez F et al 2005 Flavonoid structure-activity studies identify 6-prenylchrysin and tectochrysin as potent and specific inhibitors of breast cancer resistance protein ABCG2. Cancer Res 65:4852–4860

Chen C, Sha M, Yang S 1996 Quantitative changes of flavonoids in Epimedium koreanum Nakai in different collecting periods. Zhongguo Zhong Yao Za Zhi 21:86–88

Chen KM, Ge BF, Ma HP, Zheng RL 2004 The serum of rats administered flavonoid extract from Epimedium sagittatum but not the extract itself enhances the development of rat calvarial osteoblast-like cells in vitro. Pharmazie 59:61–64

Chen KM, Ge BF, Ma HP, Liu XY, Bai MH, Wang Y 2005 Icariin, a flavonoid from the herb Epimedium enhances the osteogenic differentiation of rat primary bone marrow stromal cells. Pharmazie 60:939–942

Critchley LA, Chen DQ, Lee A, Thomas GN, Tomlinson B 2005 A survey of Chinese herbal medicine intake amongst preoperative patients in Hong Kong. Anaesth Intensive Care 33:506–513

De Naeyer A, Pocock V, Milligan S, De Keukeleire D 2005 Estrogenic activity of a polyphenolic extract of the leaves of Epimedium brevicornum. Fitoterapia 76:35–40

DiPaola RS, Zhang H, Lambert GH et al 1998 Clinical and biologic activity of an estrogenic herbal combination (PC-SPES) in prostate cancer. N Engl J Med 339:785–791

EMEA Committee for proprietary medicinal products 2001 Note for guidance on quality of herbal medicinal products

Guo B, Xiao P 1996 [Determination of flavonoids in different parts of five epimedium plants] Zhongguo Zhong Yao Za Zhi 2:523–525, 574

Guo BL, Xiao PG 2003 [Comment on main species of herba epimedii] Zhongguo Zhong Yao Za Zhi 28:303–307

Hong DY, Lau AJ, Yeo CL et al 2005 Genetic diversity and variation of saponin contents in Panax notoginseng roots from a single farm. J Agric Food Chem 53:8460–8467

Jiang YN, Mo HY, Chen JM 2002 [Effects of epimedium total flavonoids phytosomes on preventing and treating bone-loss of ovariectomized rats] Zhongguo Zhong Yao Za Zhi 27:221–224

Kitaoka M, Kadokawa H, Sugano M et al 1998 Prenylflavonoids: a new class of non-steroidal phytoestrogen (Part 1). Isolation of 8-isopentenylnaringenin and an initial study on its structure-activity relationship. Planta Med 64:511–515

Li Shi Zhen, ca 1500. Ben Cao Gang Mu 1991 edition, Xue Yuan Publishing House Beijing

Li J, Yu S, Li T, Pang S 2002 In vitro study of the effects of Epimedium on osteoclastic bone resorption in various oral mineralized tissues. Zhonghua Kou Qiang Yi Xue Za Zhi 37:391–394

Liu FC 1984 [Effects of Epimedium sagittatum and Cistanche deserticola on DNA synthesis in 'Yang-insufficiency' animal model induced by hydroxyurea] Zhong Xi Yi Jie He Za Zhi 4:488–490

Long X, Nephew KP 2006 Fulvestrant (ICI 182,780)-dependent interacting proteins mediate immobilization and degradation of estrogen receptor-alpha. J Biol Chem 281:9607–9615

Meng FH, Li YB, Xiong ZL, Jiang ZM, Li FM 2005 Osteoblastic proliferative activity of Epimedium brevicornum Maxim. Phytomedicine 12:189–193

Milligan SR, Kalita JC, Pocock V et al 2000 The endocrine activities of 8-prenylnaringenin and related hop (*Humulus lupulus* L.) flavonoids. J Clin Endocrinol Metab 85:4912–4915

Mizuno M, Iinuma M, Tanaka T, Iwashima S, Sakakibara N 1989 Seasonal fluctuation of flavonol glycosides in Epimedium species. Yakugaku Zasshi 109:271–273

Pedro M, Lourenco CF, Cidade H, Kijjoa A, Pinto M, Nascimento MS 2006 Effects of natural prenylated flavones in the phenotypical ER (+) MCF-7 and ER (−) MDA-MB-231 human breast cancer cells. Toxicol Lett 164:24–36

Schaefer O, Humpel M, Fritzemeier KH, Bohlmann R, Schleuning WD 2003 8-Prenyl naringenin is a potent ERalpha selective phytoestrogen present in hops and beer. J Steroid Biochem Mol Biol 84:359–360

Sonneveld E, Riteco JA, Jansen HJ et al 2006 Comparison of in vitro and in vivo screening models for androgenic and estrogenic activities. Toxicol Sci 89:173–187

Sovak M, Seligson AL, Konas M et al 2002 Herbal composition PC-SPES for management of prostate cancer: identification of active principles. J Natl Cancer Inst 94:1275–1281

Sun Y, Fung KP, Leung PC, Shaw PC 2005 A phylogenetic analysis of Epimedium (Berberidaceae) based on nuclear ribosomal DNA sequences. Mol Phylogenet Evol 35:287–291

The State Pharmacopoeia Commission of P.R. China 2000 In: Pharmacopoeia of The People's Republic of China (English Edition), vol. I. Chemical Industry Press, Beijing, China, p. 107

U.S. Department of Health and Human Services, Food and Drug Administration, Center for Drug Evaluation and Research 2004 Guidance for Industry, Botanical Drug Products

Wang ZQ, Lou YJ 2004 Proliferation-stimulating effects of icaritin and desmethylicaritin in MCF-7 cells. Eur J Pharmacol 504:147–153

Wang S, Paris F, Sultan CS et al 2005 Recombinant cell ultrasensitive bioassay for measurement of estrogens in postmenopausal women. J Clin Endocrinol Metab 90:1407–1413

Wu T, Cui L, Zhang Z et al 1996 [Experimental study on antagonizing action of herba Epimedii on side effects induced by glucocorticoids] Zhongguo Zhong Yao Za Zhi 21:748–751, 763

Wu H, Lien EJ, Lien LL 2003 Chemical and pharmacological investigations of Epimedium species: a survey. Prog Drug Res 60:1–57

Yang ZZ 1985 Clinical applications of Yinyanghuo. Zhejiang J Trad Chin Med 20:478–480

Yap SP, Yong EL 2005 Epimedium species. In: Packer L, Ong CL, Halliwell B (eds) Herbal and traditional medicine: Molecular aspects of health. Marcel Dekker, New York p 229–246

Yap SP, Shen P, Butler MS, Gong Y, Loy CJ, Yong EL 2005 New estrogenic prenylflavone from
    Epimedium brevicornum inhibits the growth of breast cancer cells. Planta Med 71:
    114–119
Yin XX, Chen ZQ, Dang GT, Ma QJ, Liu ZJ 2005 [Effects of Epimedium pubescens icariine
    on proliferation and differentiation of human osteoblasts] Zhongguo Zhong Yao Za Zhi
    30:289–291
Yu S, Chen K, Li S, Zhang K 1999 In vitro and in vivo studies of the effect of a Chinese herb
    medicine on osteoclastic bone resorption. Chin J Dent Res 2:7–11
Zhang G, Qin L, Hung WY 2006 Flavonoids derived from herbal Epimedium Brevicornum
    Maxim prevent OVX-induced osteoporosis in rats independent of its enhancement in intes-
    tinal calcium absorption. Bone 38:818–825

DISCUSSION

Boobis: I wanted to ask about the area under the *Epimedium* bioactivity curve, where there was a marked difference in route of delivery. When you gave *Epimedium* subcutaneously you got a disparity between the oestrogen receptor activation and the chemical assay. But when it was given orally, the numbers seem consistent. The oral route seems to be affecting potency. Something strange is happening on subcutaneous delivery.

Yong: It could be an error. This is just our first experiment.

Boobis: I'd like also to comment on the MCF7 response. You said it was biphasic, with proliferation and then targeted destruction of the receptor. How do you envisage developing a preparation with those properties that can be used effectively in patients? If you don't get the dose just right, you will get a stimulation of proliferation at lower doses in some subjects.

Yong: It could be that oestrogenic effects are present, but because the drug itself suppresses the oestrogen receptor in a dose-dependent fashion, up to a certain stage it would have the beneficial effects of oestrogen. But you wouldn't go beyond a certain point to overstimulate the receptor and therefore result in increased breast cancer risk. This is what oestradiol, which is potent, does.

Azzi: What is the molecular basis for targeting the receptor for destruction?

Yong: There is now a lot of work on how the receptor turnover is being maintained. This is not just for steroid receptors. The activity of the receptor depends on the ligand. Even small changes in the ligand can change its activity. Some of the change of activity is due to degradation. For example, degradation of the oestrogen receptor is through proteasome coupling. For some other receptors it is through increased expression. The ligand changes the configuration of the LBD, and the ubiquitin pathway recognizes it and targets it to the proteasome pathway, where it is chewed up.

Taylor: Would this potential product be marketed as a drug or a dietary supplement?

Yong: You'd have to ask my manufacturer! I don't know how it is going to be developed. If we wanted to market it as a pharmaceutical, as a highly refined extract, then the regulatory authorities would demand intense toxicology, which would bump up the cost. The neutraceutical route is cheaper, and the product is already on the market. So what incentive is there for a manufacturer to produce a high quality product?

Przyrembel: If it were to be marketed as a food it would be a novel food in the European Community (it is not traditionally eaten) and you would have to do the toxicology you mentioned. This would not be the solution. What you wouldn't have to do is the efficacy studies, but the safety requirements would be the same as for a drug.

Shekelle: It was interesting to see all the work that has gone in to getting this compound to a clinical trial. How much effort did it take you to get from whenever you first came up with this idea to the point of doing a clinical trial? What is that time period and roughly how much did it cost? The sense I got is that this was years of work involving multiple grant applications.

Yong: It took seven years.

Shekelle: Multiply that by the number of herbal compounds, and it is a daunting task. But at least from where I am sitting, this kind of work is vitally important. I am impressed at the task that may be ahead of the herbal community.

Yong: The cost is my time, my medicinal chemists and my team. I am now spending about 70% of my labs time on this project. But this is not a true reflection of the cost. For a really good product the costs would have to include cultivation under GAP conditions. Then there is harvesting, processing and manufacturing.

Shekelle: This reinforces the opinion of people like me that we have no confidence in the herbal medicines currently on the shelves in the USA.

Taylor: The current provisions do not address these substances necessarily as drugs—where there is presumably considerable rigour in the review process—but rather herbal products may fall into categories that have different review regimes as compared to drugs.

Ernst: This is a fascinating story. Paul Shekelle has identified this as the daunting task ahead of the herbal community. I don't think so: the herbal community would block this because they would research themselves out of business. You'd be likely to end up with a non-herbal product.

Shekelle: That's not what I heard him say. He said he could have gone down the line of a purified single one of those HPLC peaks, but instead he has taken all 27 because their sum may have been greater than any of the two middle peaks, and you don't know what the interaction effects are between them. I look at this as producing a purified herbal product, but not a chemical.

Ernst: I don't think this would be a herbal product any more.

Stocker: I was particularly interested in your comment that one of the potential advantages of this is that it is a 'green' medicine. What defines 'green'?

Yong: I think there is a market for these things. Why would we want to go to the supermarket and pay a premium for organic food?

Shekelle: That is a different argument. The argument is not that the food is better, but that the production process is less damaging to the earth.

Boobis: In the UK it is both.

Yong: This applies to herbs, too.

Stocker: Is there actually a definition?

Boobis: There is no legal definition.

Taylor: In the USA, substances are categorized as drugs based on their intended use usually as stated by the manufacturer. The issue of natural versus synthetic does not come into play. If the manufacturer makes no drug-like claims for the substance or product, it can be marketed as a supplement of course assuming it meets those general qualifications as laid out in the legislation known as DHSEA.

Azzi: Is it clear that the unique target of this drug is the α-oestrogen receptor?

Yong: It's α and β.

Azzi: Are there other targets?

Yong: Definitely. This is what makes these natural preparations so complex.

Wharton: If it does have oestrogen activity, is it likely to have all the same complications as conventional HRT?

Yong: The hope is that it won't, because of its self-limiting action. It has unique effects in targeting the oestrogen receptor for destruction. If you like the model of breast cancer as a cell that has entered over-drive, this mechanism fits.

Russell: What effects would you predict on bone?

Yong: Because it has oestrogen receptor β activity I'd hope that it would have activity. It's not so simple to look at because there isn't a good osteoporosis model.

Manach: Do you think that *Epimedium* consumption should be restricted to post-menopausal women?

Yong: That's a good question, because this herb is marketed as a male viagra, yet it's oestrogenic, not androgenic. I don't know.

Manach: What about for young women?

Yong: It is recommended for them, and for males too.

Aggett: How well documented are these effects of *Epimedium*? Are there well designed trials?

Yong: The short answer is no.

Aggett: In a way you were being slightly speculative, then, when you started this seven years' work.

Yong: I wasn't a believer in herbs, but one of my postdocs was. They wanted to try this. Extracts are interesting as they have effects on biological systems, and as a biologist I am interested in effects. The basic science of natural products chemistry is increasing all the time. Compounds and bioactive constituents in herbs are constantly being discovered and characterized. The number of biological pathways affected by these compounds is mind-boggling. Thus from a scientific point of view, these compounds and their parent botanicals are exciting. Besides there is always the hope that ancient cures may turn out to be effective modern medicines.

Manach: Bioavailability must be taken into account. Most of the arguments for the mechanisms of action you made are based on *in vitro* activity. There may be major compounds present in the extracts that are not absorbed at all. These will have an effect in *in vitro* studies and we might not see the same effects *in vitro*.

Communication between science and management

Hildegard Przyrembel

Federal Institute for Risk Assessment, Thielallee 88-92, D-14195 Berlin, Germany

Abstract. Four methods of communication between science and management are conceivable: (1) authoritative science, which provides results without explanation or justification; (2) authoritative management with imposition of a preconceived management decision on science, which destroys independence and credibility of the scientist; (3) usurpation or mutual invasion of both science and management into each other's territory, which is detrimental to the integrity of both; and (4) interaction between scientist and manager, in which the different tasks of both are recognized and respected. For the latter it is important to accept that the commitment of the scientist is to science only and that managers are committed to other considerations besides science. The role of the scientist is easier because of their commitment, and the results of their work are less likely to be questioned, when they succeed in communicating their working methods, treatment of (missing) data, deductions, and results in a comprehensible and logical form. The manager, however, will be in a difficult position, if these results take the form of advice which, for whatever reasons, they cannot follow. Managers may be tempted to avoid advice or, if unavoidable, to doubt its correctness or to modify its meaning, instead of justifying their decision. The major problem in communication between science and management is probably in semantics: the wording of the task of the scientist and of the scientific result in unambiguous language which is understood by the assessor and the manager, respectively.

2007 Dietary supplements and health. Wiley, Chichester (Novartis Foundation Symposium 282) p 192–201

Science has always played a decisive but often poorly defined role in the making of decisions by managers/regulators. In 1995 the Codex Alimentarius Commission formulated Statements of Principle concerning the role of science in the Codex decision making process and the extent to which other factors are taken into account. (1) The principle of sound scientific analysis and evidence was declared to be the basis for food standards, guidelines and recommendations. This would include a thorough review of all relevant information. The aim of this first statement was to assure the quality and safety of the food supply. (2) The second statement requests that '*where appropriate*' in deciding on food standards, regard should be given to other legitimate factors relevant for the health protection of consumers and for the promotion of fair practices in food trade, clearly a description of the role and the commitment of managers/regulators. The consideration of other factors besides science into the decision-making process can lead to the need for additional assessment by science.

The Codex Alimentarius Commission has clarified the roles of science and risk managers in 2001 in the Criteria for the Consideration of the Other Factors Referred to in the Second Statement of Principle. While the role of science and of food safety risk assessment are observed, other legitimate factors may be identified in the risk management process and can affect the selection of risk management options but not—and this is important—the scientific basis of risk analysis. These other factors need to be clearly documented as well as the rationale of their integration into the decision-making process (Codex Alimentarius Procedural Manual 2005).

The Working Principle for Risk Analysis for Application in the Framework of the Codex Alimentarius was adopted by the Codex Alimentarius Commission in 2003 (CAC Procedural Manual 2005). These working principles contain descriptions of the three components of risk analysis—risk assessment, risk management and risk communication—and, while strongly recommending the functional separation of the risk assessment and risk management, underline the importance of an interaction between risk managers and risk assessors and provide rules for such interaction. Desirable communication between risk assessors and risk managers is described as 'clear, interactive and documented' and it is set apart from the task of communication with and informing of the public and of interested parties, a task that belongs to both assessors and managers. As usual, the reality is not always in line with expectations, and communication between risk assessors (or science) and risk managers can be difficult.

Communication is a science and there are experts in this field. The author of this text is not such an expert, but has some experience in the difficulties of communication between science and management and will describe these difficulties using examples from her own tasks as a scientific assessor of foods, food ingredients and nutrients for her government and for the Scientific Committee for Food of the European Commission and for the European Food Safety Authority (EFSA).

Effective communication between science and management

The Working Principles for Risk Analysis for Application in the Framework of the Codex Alimentarius have been chosen as an international reference text for the communicative obligations of science and management, while acknowledging that other national and international institutions have provided similar texts. These texts deal with science and scientists that are in direct contact with managers in the assessment of risks and the evaluation of scientific data. Emphasis on communicative activities in the citations has been added. From the citations the word 'risk' can be deleted for the following discussion of possible ways of communication between science and management, because this communication does not only concern risks.

> '*Effective communication and consultation* with all interested parties should be ensured (throughout the risk analysis . . .
>
> There should be a functional separation of risk assessment and risk management, in order to ensure the scientific integrity of the risk assessment, to avoid confusion over the functions to be performed by risk assessors and risk managers and to reduce any conflict of interest. However, it is recognised that risk analysis is an iterative process, and **interaction between** risk **managers and** risk **assessors is essential** . . .
>
> Risk assessment policy should be established by risk managers in advance of risk assessment, **in consultation** with risk assessors and all other interested parties . . .
>
> The mandate given by risk managers to risk assessors should be as clear as possible.
>
> The conclusion of the risk assessment including a risk estimate, if available, should be presented **in a readily understandable and useful form** to risk managers and made available to other risk assessors and interested parties so that they can review the assessment.
>
> Risk management should follow a structured approach, . . . evaluation of risk management options, monitoring and review of the decisions taken. The decisions should be based on risk assessment . . .'
>
> (Codex Alimentarius Commission, Procedural Manual 2005)

Ideally scientists and managers should communicate in an interactive manner, that is, the pathway of communication is passable in both directions and at both ends somebody is listening, understanding and answering. However, even if the pathway is passable in both directions, reception and understanding of the message may be disturbed. On the side of the manager this may be due to the fact, that he has already made a decision and/or does not want a contribution from science. On the side of the scientist this may be caused by misunderstanding the task and/or by providing his assessment in unintelligible language or in a form which is not useful. If in such cases of a disturbed or non-existing communication both sides persist in communicating with the public or with other interested parties the result is likely to be divergent messages with only a small chance for apparent consent.

There are different reasons for such disturbances in the communication between science and managers, which shall be named:

1. Authoritative science
2. Authoritative management
3. Usurpation of the task of the partner or confusion of tasks, as opposed to the desirable situation of
4. Interaction in mutual respect and consideration.

Authoritative science

Authoritative science means that the scientific assessor, although fulfilling a task appointed to him or her, determines the scope, extent and intensity of his assessment, and also the form of its result, preferably as a single advice for action to be performed by the manager. I remember the glorious days for the assessor: in the Federal Institute for Health we used to write one-page reports of the format, '*We know what is good or bad, necessary or unnecessary, useful or useless, therefore, a specified action should be taken*'. Sometimes we were kind enough to explain and to refer to scientific data.

The following example is an illustration of that procedure. When asked more than 15 years ago if the addition of L-carnitine (100 mg/L) to a sports drink was nutritionally sound and acceptable, we replied that it was not, because no increase in physical performance could be expected, and if it were possible that such an effect could be achieved, the proposed amount of carnitine to be added would be too small. Moreover, there were no losses of carnitine due to exercise which could not be compensated by endogenous synthesis. Proposed claims on an enhanced physical performance were misleading claims. Carnitine should be reserved for foods for particular nutritional purposes. No risk assessment was performed. The product was not marketed.

Of course, this position was overruled by developments in the European Community and elsewhere, when risk analysis became en-vogue and a prerequisite for management decisions. We had to perform a risk assessment for L-carnitine and found it to be well tolerated and unlikely to cause severe adverse effects on health even when consumed in amounts much higher than possible from normal diets. Even then we tried to reassert ourselves by stating that such intakes were not necessary and, without success, prevent the widespread use of carnitine in fortified food and food supplements.

As an aside, for a nutrition-science minded person, the reliance of management decisions solely on the confirmed unlikelihood of adverse effects, i.e. on toxicological considerations, can pose a certain problem for the concept of nutrition as a science and, therefore, of self-esteem.

Conclusion

Science should respond to the point and not to unasked questions. If science responds to unasked questions it should draw attention to the fact and explain the necessity to do so. Science must perform scientific assessments, which allow the manager to follow the deliberations leading to the conclusion and which include a description of the uncertainties in the evaluation.

Authoritative management

This situation can occur when the manager knows clearly what he will decide and tries to influence the assessment process accordingly. Although the decision is made, a (pseudo)-scientific justification is needed, because it will make communication of the decision to the public easier and is expected to minimize questioning of the decision. Assessors should be aware of such pitfalls and avoid them if possible, to preserve their integrity. A variant of this is the situation when the management has already decided on an action on the advice of the scientific assessor but has not yet acted, and when the assessor reconsiders and modifies his assessment because of more recent or additional data having become available. Because legislative procedures take time and are cumbersome, the management might be tempted to disregard the updated assessment or even to prevent its publication.

The most subtle manner for the management to exert its authority is in formulating the request for assessment, called the terms of reference. As an example, the terms of reference given by the European Commission to the European Food Safety Authority (EFSA) relating to tolerable upper intake levels (UL) of vitamins and minerals shall be given: '. . . *the European Food Safety Authority is asked 1) to review the upper levels of daily intakes that are unlikely to pose a risk of adverse health effects; 2) to provide the basis for the establishment of safety factors, where necessary, which would ensure the safety of fortified foods and food supplements containing the aforementioned nutrients*'. No. (1) has been confidently interpreted by the EFSA to mean that it should review and evaluate all available data and perform a risk-assessment of the nutrient under consideration according to established procedures and define a tolerable UL, if the available data are sufficient, and/or provide a risk characterization of the nutrient with regard to the observed or potential intake levels in the population, and the probability of the occurrence and of the severity of possible adverse effects on health. The question remains open if this result provides the basis for the establishment of the safety factors mentioned in No. (2) of the terms of reference, which are clearly not to be established by EFSA, especially as the nature of these safety factors is not defined. They are not mentioned specifically in the European directive on food supplements (Directive 2002/46/EC). How are the statements found in the risk characterization section of some assessment reports, where the likelihood for adverse effects or for exceeding the UL is low—which by itself are two different things—to be a basis for the establishment of safety factors? Should EFSA have commented on this or discussed the impact of different options?

Conclusion

Managers should define the task of the assessors clearly, and assessors should be aware of the intentions of the managers in order to produce the adequate basis for management decisions. If the request to assessors is ambiguous, clarification should be asked for. While managers should be informed about the process of the assessment they should not influence the result.

Usurpation of the task of the partner or confusion of tasks

As indicated, it is always tempting to perform the task of another, either because one is convinced that one knows best what to do or to avoid interference. Since the functional and physical separation of risk assessment and risk management in Germany in two institutions it has become easier for assessors to concentrate on the assessment and not bother about concrete advice for action by the manager. This also ensures that the assessors refrain from shaping their assessment according to their ideas of what the management should decide, and to accept that other factors besides their assessment will influence a management decision. After a period of accustomization for particularly the assessors–including consistent wording and structure of the assessment reports–communication is flourishing, possible options for management decisions are discussed and responsibilities are clearly divided.

A more insidious form of transgression can still occur from the side of the management in deliberate or unintentional mis- or overinterpretation of the scientific assessment. An example is the use of the assessment done by EFSA on the UL of boron, and on boric acid and sodium borate as nutrient sources of boron (EFSA 2004a,b). In the first assessment a UL of 10 mg boron/day for adults was set, in the second report boric acid and sodium borate were considered to be *'suitable for use in foods for particular nutritional uses, food supplements and foods intended for the general population providing the UL is not exceeded'*. This statement was the basis for draft proposals for the amendment of the directives on substances which may be added to foods for particular nutritional purposes (Directive 2001/15/EC) and on food supplements (Directive 2002/46/EC) to include both boric acid and sodium borate as sources of boron because EFSA had provided *'a favourable scientific evaluation'*. Some member countries who disagreed with this proposed action, argued that the assessment by EFSA did not include an evaluation of the impact of this action and the 'favourable' opinion of EFSA concerned only the suitability of both substances as sources of boron in foods and not the decision to permit the addition. An assessment of the likelihood of exceeding the UL for boron from all sources, both dietary and environmental, of the impact of addition of boron to foods on that likelihood, and of the probability and extent of the occurrence of adverse effects as a consequence of additions of boron sources to food has yet to be performed.

Conclusion

Both assessors and managers should concentrate on performing the task for which they are responsible. If the assessment does not include the evaluation of the impact of management decisions, either the decision should not be justified with the assessment or the assessors should be requested to assess the impact of that decision.

Summary and conclusions

The description of suboptimal forms of communication between science and management allows us to formulate requirements for desirable and effective forms, of which the first is that there should be interactive communication, meaning that results of scientific assessment and management decisions are not just transferred to the manager and scientist, respectively, but are discussed together.

The tasks of the scientific assessor and manager are kept functionally separate and are mutually recognised and respected. There is time for regular contact and interchange.

The tasks for the scientific assessor are clearly formulated, preferably in co-operation between managers and assessors. If they are ambiguous, assessors must request unambiguous reformulation.

The result of the assessment is formulated unambiguously and covers all aspects of the terms of reference. Managers must request additional assessment or rewording if the assessment is incomplete or open to misinterpretation.

Assessors refrain from anticipating management decisions.

Managers refrain from use of the assessment to justify management decisions that were not part of the assessment.

The actuality of an existing assessment should be controlled by both the assessor and the manager, and both can initiate a re-assessment.

Both assessors and managers respect each other, have patience and develop a common language.

References

Codex Alimentarius Commission (CAC) Procedural Manual 2005 14th edn ISSN 1020–8070
Commission Directive 2001/15/EC of 15 February 2001 on substances that may be added for specific nutritional purposes in foods for particular nutritional uses. OJEC L52/19–25
Directive 2002/46/EC of the European Parliament and of the Council of 10 June 2002 on the approximation of the laws of the Member States relating to food supplements. OJEC L182/51–57
EFSA 2004a Opinion of the Scientific Panel on Dietetic Products, Nutrition and Allergies on a request from the Commission related to the Tolerable Upper Intake Level of Boron (Sodium Borate and Boric Acid). EFSA J 80:1–22
EFSA 2004b Statement of the scientific panel on food additives, flavourings, processing aids and materials in contact with food on a request from the commission related to boric acid and sodium borate as nutrient sources of boron. 08.12.2004

DISCUSSION

Boobis: The issue of clarifying what is required of the risk assessor is an important one. Experience has shown that it is a two-way problem: the risk assessor shouldn't presume to give information the risk manager hasn't asked for, but the difficulty is that sometimes the risk manager hasn't asked the right question. This is where two-way communication is really important. At EFSA we have discovered that often we have to discuss with the risk managers what they are trying to get out of us to ensure that the question formulated will give them the answer they need to perform the risk management function.

Przyrembel: One of the best examples is the fish report from EFSA. I was in a working group for this report and we spent some days deciding what the commission wanted to know. In the end we still didn't deliver a very clear answer.

Aggett: I think that putting in the issue about problem formulation and emphasizing the iterative nature of developing that question is helpful. We are getting better at it. What is needed is that we are well instructed. Those of us who are scientists come in and we are not fully aware of the problems. Now we are starting to work better with managers we are starting to realize the pressures that they are under from politicians. This is becoming quite apparent in our agencies.

Przyrembel: I know they are under pressure, but I would sometimes like an explanation of this pressure. It is difficult if you are told that there is no other way and that the result must be like this, and, therefore, you are requested to come to this result. This is not honest. Managers should explain why they are forced or are under pressure to come to this result. I don't think there is anything dishonourable about this.

Scott: You took the example of boron. It seemed that the question the risk assessors were asked is whether they thought boron is necessary, but they gave as part of their answer that, 'Yes, you can add boron but you mustn't exceed the UL'. Then you could say to managers it is up to them to figure out how they don't exceed the UL: it is a management issue. Is that what happened?

Przyrembel: I think the wording they chose in the draft of the directive was incorrect. This emphasizes how important the choice of language is. If you ask one EFSA panel to provide a UL and another EFSA panel to give an opinion if these two substances are good sources, then you can't say that EFSA gave a favourable statement about the addition of these substances to these kinds of foods, because EFSA was not asked for its opinion on this management decision.

Taylor: I wanted to address the outcome (or question) relevant to L-carnitine as listed on your slide. Another 'communication' issue is obtaining clarity about whose task it is to answer certain questions. Inadvertently, certain questions can get passed to the 'wrong' parties, so to speak, for many reasons including a desire to avoid responsibility. During the FAO/WHO risk assessment workshop, the participants looked at the nature of the questions asked of nutrient risk assessors, and at times it seemed that questions more appropriate for managers were being assigned to assessors. On the slide in your presentation, the very first question relative to the L-carnitine assessment was, 'Is the issue of L-carnitine. . . . and therefore acceptable?' It would seem to many that the question of 'and therefore acceptable' is a one for risk managers not for risk assessors.

Przyrembel: This was only an example of how, in this case, the assessor did not respond correctly.

Taylor: It could be argued that if there was a back and forth through problem formulation, you might have been able as a risk assessor to push this back on the risk manager. Isn't that their domain?

Aggett: In my experience, it is only recently that panels have realized that they can or should do that. I gave the example earlier about the deferred approval for adding ω-3 fatty acids to infant formula for term infants. The novel foods panel decided to reject their addition to a formula for term infants. In retrospect, this was an inappropriate reaction because it was not what that panel was being asked to do. If there had been some proper interaction between advisory processes spanning nutrition, safety and innovations then a clearer decision would have been made relevant to what was being asked, and the residual decision about whether they should have been there at all could have been decided by another agency. That there wasn't such an agency did influence the decision, but by and large it is a good example of a panel not sticking to the question. Fortunately, the new advisory structure in a single Food Standards Agency should avoid such problems in the future.

Przyrembel: A speciality of EFSA is that they have different scientific panels. It is another speciality that sometimes a task is given to what looks like the wrong panel. The case of the infant formula is a good example of this. It should have gone to the nutrition panel.

Alexander: In my experience, when we run into trouble it's almost always with the problem formulation. You mentioned the fish report, which I was a part of. The question originally came from the European parliament, and was very vague. Both the Commission and EFSA had problems with interpretation of what that really meant. And we could not ask since it was coming from an outside body.

Przyrembel: This is the only report EFSA has ever produced which starts with an interpretation of the terms of reference and a statement to which questions EFSA will respond.

Stocker: I have no direct experience of these issues, but it seems that they appear in many facets. Language is important. Another important aspect is the knowledge base. In some examples that you reported that the miscommunication (or lack of clear communication) may have to do with a difference in knowledge base between management and assessors. I wonder whether the knowledge base of management is such that sometimes they are simply not able to ask appropriate questions because they lack the appropriate scientific knowledge. A more common knowledge base of management and assessors may help facilitate effective communication.

Przyrembel: The European Commission is in direct verbal contact with the assessors and this is improving. The first thing we do is discuss the scope and extent of the opinion to be produced with the representatives of the managers. If we can't agree on this, the representatives go back and the question is reformulated. There is already an exchange of knowledge.

Yetley: The principles you discussed, in terms of the interface and the appropriate roles of the scientists versus the user, have broader applicability. You want the evidence-based reviews to appropriately address the questions that have been asked, but not to make conclusions that go beyond their focus and terms of reference. For example, an evidence-based review could describe the strengths and weaknesses of the evidence relative to a particular question; subsequently, the review would then be used by an advisory committee or government agency in a policy setting to identify research priorities. In many cases, science and policy decisions need to be independently determined. Policy decisions differ from scientific evaluations in that they need to take into account many factors in addition to the relevant science (e.g. costs); however, they benefit from being fully informed by the science. Conversely, the independence of a scientific evaluation is essential to maintaining the integrity of the scientific process and results, and to avoid accusations—whether fair or unfair—of having manipulated the science to unduly influence a policy outcome.

Dietary supplements and health: the research agenda

Paul M. Coates

Office of Dietary Supplements, National Institutes of Health, 6100 Executive Boulevard, Suite 3B01, Bethesda, MD 20892-7517, USA

Abstract. Research needs to evaluate the role of dietary supplements in human health abound, yet funds to support all of the possible opportunities do not. Government agencies, such as the National Institutes of Health (NIH) in the USA, remain the chief sponsors of research in this area. They face the challenge of competing priorities, such as critical disease-oriented research, basic biomedical and technological development, and prevention-related research. Dietary supplements are widely used for health promotion and disease prevention, sometimes with minimal science to support their use. There is a need for focused research efforts to better address issues of efficacy, safety and quality of dietary supplements. At the same time, fundamental studies of their mechanisms of action are needed. In addition, resources to support research in this area are required: on the one hand, basic tools (analytical methods, characterization of ingredients) need to be developed and validated, and on the other, tools to understand patterns of supplement use in populations and study designs to assess their efficacy and safety need refining. These efforts benefit greatly from partnerships among government agencies and with the academic and private sectors.

2007 Dietary supplements and health. Wiley, Chichester (Novartis Foundation Symposium 282) p 202–211

Consumers are increasingly trying to take charge of their own healthcare, particularly in environments such as the USA where concerns have been raised about the rising costs and the depersonalization of healthcare services. Lifestyle interventions aimed at promoting health and reducing the risk of chronic disease have gained attention; these include attempts to improve dietary habits, to increase physical activity, to curtail risky behaviours, and to increase consumption of dietary supplements. In the last example, consumers use dietary supplements because they may perceive the need to make up for nutritional inadequacies, to promote and maintain health, to serve as a kind of insurance policy against the functional declines seen with ageing, or to manage and sometimes treat diseases. The Office of Dietary Supplements (ODS) was established at the National Institutes of Health

(NIH) as a result of the Dietary Supplement Health and Education Act of 1994, with a mission to enhance and support research on dietary supplements that would ultimately be translated into effective information for consumers. The dietary supplement category in the US marketplace embraces a wide variety of products, including vitamins, minerals and amino acids, but also herbs and other botanicals, as well as other products that consumers use to meet the needs noted above.

What factors enter into a discussion of research priorities?

In the case of some dietary supplement ingredients, a great deal of information is available, while for others, the data set is meagre or immature. More is known, in general, about the biological effects of micronutrients than about the effects of herbs. In many respects, however, the scientific needs are quite comparable. Issues of benefit, risk, and product reliability dominate the discussion.

- What are the optimum research designs that will get at the health effects of supplements? Sometimes it is appropriate to run clinical trials and the NIH has funded many that have used dietary supplements as interventions, e.g. selenium and vitamin E to prevent prostate cancer, St. John's wort to prevent minor depression (the complete list of ongoing NIH clinical trials can be viewed at *http://clinicaltrials.gov*). On the other hand, the complexity (and therefore the cost and difficulty) of running trials with interventions that are readily available to subjects can be prohibitive. If the effect is relatively small and perhaps only realized over a long period of exposure—as would likely be the case with supplements—then there is a need for careful prioritization and planning of trials with these agents.
- How does one assess effects of micronutrients when their background intake from foods has not been taken into account? This means that effective, validated tools for dietary assessment must be in place. The lack of reliable indicators of dietary intake in trials is one of the factors that repeatedly surfaces in systematic reviews of dietary supplement efficacy.
- Does one impute safety from short-term clinical trials where no adverse events are identified? If health benefits are only likely to be recognized after long-term exposure, is it reasonable to expect that harms will only occur in the short term?
- Has the clinical trial of an herb used a product that is in the marketplace? If so, then the results of that trial may be applicable only to that product.
- Has the product been well-characterized and is it stable and reliable for the duration of the study? What endpoints are appropriate to study? Since dietary supplements in the USA are legally marketed only for health promotion and for reducing the risk of chronic disease, is it legitimate to evaluate their health

effects against disease endpoints? In practice, how do you measure 'health promotion'?
• Disease endpoints may take years to develop, so what about measuring effects against biomarkers of disease? How have these biomarkers been validated?

Research agenda development in a public agency

The NIH has developed a long-term strategy for research priority development called the NIH Roadmap Initiative. It contains several featured areas that have been recognized as crucial (*www.nihroadmap.nih.gov*) and highlights strategies to accomplish them: New Pathways to Discovery, Research Teams of the Future, and Re-engineering the Clinical Research Enterprise. All are relevant to the efforts related to dietary supplements and health. To give a few examples that ODS currently employs in efforts to enhance research related to dietary supplements:

• Building interdisciplinary research teams that focus on botanical dietary supplements (*http://dietary-supplements.info.nih.gov/Research/Dietary_Supplement_Research_Centers.aspx*)
• Co-funding research with NIH Institutes and Centers on basic and clinical research (*http://dietary-supplements.info.nih.gov/Funding/Grants_Contracts.aspx*)
• Developing unique resources for dietary supplement research (*http://dietary-supplements.info.nih.gov/Research/research.aspx*)

The ODS Strategic Plan for 2004–2009 (*http://dietary-supplements.info.nih.gov/Strategic_Planning_2004–2009/Planning.aspx*) placed major emphasis on research related to disease prevention, but at the same time recognized the continuing need to build basic research capacity and to capitalize on the existing resources of other agencies in order to support all areas of science related to dietary supplements. There are examples of emerging initiatives with other NIH units (notably the National Cancer Institute and the National Center for Complementary and Alternative Medicine [NCCAM]) and other Federal health organizations (for example, the Food and Drug Administration [FDA], the Centers for Disease Control and Prevention [CDC], the Agency for Healthcare Research and Quality [AHRQ], and the Office of Disease Prevention and Health Promotion—all in the US Department of Health and Human Services) that bear discussion.

One is related to a critical evaluation of the role of biomarkers in nutritional interventions for chronic diseases such as cancer. The NCI, FDA and ODS jointly sponsored a workshop on this issue in 2005, the conclusions of which can be found on the NCI web site at *http://www.cancer.gov/prevention/Workshop_Proceedings.doc*. Among the important outcomes were: recognition of the need to expand the use of relevant animal models, particularly to assess nutritional interventions; the need for biomarker discovery and validation, particularly in the area of disease prevention; recognition that a battery of biomarkers may be needed to provide a potential 'fingerprint' of disease; the need for standardization of laboratory procedures and verification of findings to improve reproducibility. At this stage, there are few biomarkers that predict the effect of nutritional interventions (including dietary supplements) on chronic disease outcomes. The discovery and validation of these will be crucial to the development of appropriate clinical trial designs in the future.

Evidence-based review of efficacy and safety

The ODS has employed the principles of evidence-based medicine, including systematic review and appropriate analytical tools such as meta-analysis, to evaluate dietary supplement efficacy and safety. The principal reason for this is to have an objective, transparent mechanism for evaluation that will assist ODS and its partners in setting future research agendas. Since this programme began, ODS has sponsored systematic reviews, often in partnership with other NIH entities, of:

- Chromium and insulin action
- Omega-3 fatty acids in prevention of heart disease and other conditions
- Health effects of soy
- Ephedra in weight loss and athletic performance
- Antioxidants in berries and neurodegenerative diseases.

A complete list of these and other related systematic reviews of dietary supplement effects can be found on the AHRQ website (*www.ahrq.gov*), with whom we have partnered in this effort.

Using chromium as an example of the journey from concept to research, ODS convened a small workshop in 1999 to review research needs in the area of chromium and insulin action. The workshop concluded that while there was some evidence to support a role for chromium in potentiating insulin action, there was little research that directly pointed to its role in reducing the risk of developing type 2 diabetes mellitus. ODS, along with the NCCAM and the National Institute of Diabetes and Digestive and Kidney Diseases (NIDDK), sponsored an evidence report to systematically review the literature on several questions about research related to chromium and insulin action (Althuis et al 2002). The report confirmed and extended the workshop findings and pointed to several areas of research that needed to be addressed. ODS, NCCAM and NIDDK issued an announcement declaring its interest in supporting research that addressed these needs and over the ensuing several years, approximately a dozen grants were funded that focused on methodological issues (measurement, model development) and both preclinical and early-phase clinical research. As a result, there are much better tools in place to take the next steps, namely to perform large-scale clinical trials.

This logical progression of ideas from concept to implementation is not always employed when it comes to assessing the health effects of dietary supplements—or for that matter, the putative health-promoting effects of some of these same compounds in foods. It may well be the reason why there is often suspicion about the nature of these health effects: too frequently, the reasoning is shaky or the science may be at an immature stage.

There is a countervailing opinion that even if the science is not solid, there may be compelling reasons for ensuring that consumers have access to health-promoting strategies such as dietary supplements. If there is no evidence of harm, why not encourage a behaviour that sustains health? Furthermore, given the challenges to performing good clinical research in this area as noted above, it may never be possible to conduct the definitive clinical trial for any single dietary supplement, let alone the whole array of products in the marketplace.

Research resources: analytical tools and databases

In addition to supporting research through competitive grant mechanisms, ODS has recognized the importance of building necessary research resources to meet scientific needs in this area. These include:

- Establishing a programme for the development, validation, and dissemination of analytical methods and reference standards for dietary supplement ingredients.
- Databases of dietary supplements for use in assessing intake of dietary supplements in epidemiological studies.

Outcomes and conclusions

All of these research and related efforts ultimately bear on the development of recommendations for the public. They find their way into the development of, for example, the Dietary Reference Intakes (DRIs), a body of detailed dietary intake recommendations regularly reviewed and updated by the Food and Nutrition Board of the National Academies of Science in the USA. The DRIs, in turn, form the basis for national food and nutrition policy such as the Dietary Guidelines for Americans.

These research efforts also help ODS to fulfil its requirement to provide effective information for consumers as they make healthcare choices that include the use of dietary supplements. To that end, ODS publishes a series of consumer-oriented Fact Sheets that summarize current information about dietary supplements (*http://*

*ods.od.nih.gov/Health_Information/Information_About_Individual_Dietary_*
Supplements.aspx).

Discovery in this area will need to occur on a number of related fronts, given the fact that active components of dietary supplements, as well as foods, likely have multiple targets and health effects (Coates & Milner 2006). Linkages among dietary habits, supplement use and overall health emanate from a multitude of epidemiological, preclinical and clinical studies. Unfortunately, inconsistencies arising from these investigations have sometimes clouded our understanding of the health effects of these interventions. These inconsistencies result in part from variations in experimental designs that have been used to address nutritional interventions for health outcomes, but they also reflect the multifactorial and complex nature of health, individual variability, and the interactions among dietary constituents when they are consumed in foods or individually as dietary supplements. Tools are emerging to better address genomic, proteomic, and metabolomic aspects of dietary interventions that will shed light on their health effects, including why variation in response may arise. This will guide future research that will, in turn, provide a sounder basis for making public health recommendations.

References

Althuis MD, Jordan NE, Ludington EA, Wittes JT 2002 Glucose and insulin responses to dietary chromium supplements: a meta-analysis. Am J Clin Nutr 76:148–155
Coates PM, Milner JA 2006 Bioactive components of foods and dietary supplements. In: Bowman B, Russell R (eds) Present knowledge in nutrition, 9th edition, ILSI Press, Washington DC p 959–967

DISCUSSION

Scott: One of the things that strikes me about research on dietary supplements is that most of the studies are underpowered. They create this rather repetitive situation where they produce negative results. It is true that you can't prove a negative, but you can say things like, 'if this does have an effect, it is below a certain level'. You simply can't make an application to the Food Standards Agency (FSA) in the UK without a statistician being involved. This has streamlined a lot of thinking. You don't need a lot of power to look for differences in biomarkers, but to look for disease effects you need huge power, unless you are hoping this is going to be, for example, 50% of the cause of colon cancer.

Coates: These are issues we think about. If anything, we would want to be able to use the evidence analysis to be able to inform future major clinical trials. With omega-3 fatty acids, one of the issues is that we don't expect that there will be a huge effect. This is a comfort in some respects, because the chance of having a huge positive effect increases the potential that some people may suffer harm. We have this challenge: where the effects sizes are small, they may not be realized over a small time. This causes the costs of clinical investigation to increase. In planning for a trial, I don't think that any NIH Institute can afford to invest in something where there hasn't been a considerable amount of thought given to issues like power and effect size. We do want to try to make these studies more efficient, so if biomarkers can be validated, we would use them. Currently we don't have any that we can use with confidence. It would be very nice, for example, to design trials where we knew something in advance about the possibility of a response being associated with a particular genotype or haplotype.

Shekelle: This issue is not related solely to dietary supplement studies. The issue of power is one that comes up a lot in intervention studies. Usually, the lack of power for a major study is not because the investigators didn't do a good job of planning the trial, but rather because if there haven't been big trials you are usually powering on a study for effect size that is based on smaller trials or observational data. There is a well established relationship: effect size usually declines as studies get bigger and better done. The other thing that has plagued a fair number of cardiovascular studies is that the placebo or conventional treatment group is getting better over time. By the time a trial is carried out, the event rate in the control group may have reduced significantly. Institute or investigator competence is rarely the problem. This is opposed to pharmaceutical industry-funded studies, where the problem is deliberate powering and underpowering in order to find what you are trying to find.

Russell: One of the issues I keep returning to in dietary supplement research is the dosage effects and getting a handle on intermediate responses between something that may happen at a high dose versus a dietary dose. What we saw in the β-carotene situation is that the high-dose response was just the opposite of the low dose response, with regard to cancer. We need some thought about having some kind of dose calibrations in between a massive pharmacological dose (which people tend to use because they think they are going to get the best effect that way) and a dietary dose.

Coates: I think this is crucial. One of the reasons that I'm glad that Beth Yetley is in the Office of Dietary Supplements is that she never stops reminding us about this issue of dosing and how important it is to have this embedded into the early stages of development. Part of the problem is that dietary supplements are available in the marketplace. In some cases we are back-filling the research that wasn't done. Some of it will be hard to do, but if we set the standard that there are some fundamental questions and tools that need to be in place, we have a better chance of being able to get realistic assessments of the efficacy and safety of these products.

Boobis: To what extent do you enter into partnership with the producers? Someone mentioned that the original ATBC (α-tocopherol/β-carotene) trial was in partnership. As an independent looking from the outside in, all this work has been done on the public purse, but the people who stand to make money are the supplements manufacturers. Shouldn't they be contributing to the cost?

Coates: That is a good point. We couldn't fund large clinical trials from our budget, so we do all of these sorts of things in partnership. NCCAM has had to deal with some of the vexing issues of recruiting potential suppliers and the like.

Klein: NCCAM does not seek funds from industry to directly support a study. NIH is concerned about conflict (or appearances of conflict) of interest. It is not uncommon that industry would want to be involved in protocol development if it also provided financial support for the study. This can be perceived as a conflict. Therefore, to avoid conflict of interest (or its appearances) NCCAM will not seek direct funding from industry for a study but will seek donation of product. Selection of product is usually a competitive process so that selection is made on an objective, scientific basis.

Aggett: I suppose one of the key points is the issue of quality and the use of biomarkers that are fit for purpose. One of the ways of helping progress things might be to look at whether there could be advantages in having some certified outcomes. These could be important outcomes that have been identified by regulatory bodies in consultation with practitioners of the key issues. An example could be protein expression in the early pathogenesis of cancer. I don't think we'll come up with many, and some of the most useful will be old measurements such as blood pressure.

Coates: I appreciate the suggestion. This kind of thinking is necessary.

Scott: There is quite a significant spend on botanicals or herbals. What is the motivation for this? One could be saying that here is something that is 3000 years old, and there probably is an active substance present, as we heard earlier. But then we also heard earlier that even if we did purify out an active ingredient, this would then be a medicine and the herbal people don't want to do this. One of the reasons people want to use these herbal remedies is that they are 'natural'. For me, the important thing is to make sure they don't do any harm, but you are looking for efficacy, and it is unlikely you'll be able to demonstrate this with multi-component preparations whose mechanism of action is unknown. Perhaps the agenda is that while you are looking for efficacy, safety is driving what you do. Is that correct?

Coates: Another piece of congressional language called on the ODS to create a series of multidisciplinary botanical research centres around the USA. This was very much in NCCAM's best interest as well, so we partnered with them in funding these centres.

Russell: There should be a standardized way of looking at botanicals, characterizing them so we know what we are dealing with. We have heard that this can't just involve chemical analysis, but also requires a standardized bioassay. Otherwise we will end up with a mass of data confounded by the fact that different products are being researched under the same name.

Coates: It's fair to say that this is the starting point for a lot of things. For some products or ingredients that are better characterized, this should be the model. It is quite an investment in time and resources. We have to be careful about how we set our priorities or we could be throwing money away for generations. NCCAM and we have some similar goals, but we don't overlap entirely. Ultimately, I know what I would like to see happen: that there is an appropriate evaluation of these ingredients as dietary supplements, i.e. for their effects in health promotion and for reducing the risk of chronic diseases.

Boobis: I don't think we have a gold standard yet for characterization of these products. If we take the example of the cell-based bioassay, I have a suspicion that if you took the same dozen products and put them through an *in vivo* assay for outcome, you would get a different ranking. The assumption that they are all going to work through oestrogen receptor activation is probably incorrect. It is a difficult area: we need to discuss how we will break this problem. It is a key issue and we don't have the answer. Taxonomy won't work; chemical profiling won't work. For complex mixtures, simple *in vitro* bioassays won't work either.

Yetley: You said proper characterization was a gold standard, but isn't it rather an essential piece?

Russell: You are right. We need an approach like a bioassay which everyone agrees on, at the start.

Yong: I used this assay because this is my research interest. But I could imagine in these days of different cell lines and DNA arrays, we could come up with a few standardized platforms, with one cell line for toxicology, another for efficacy and a microarray approach to examine a whole series of gene changes.

Boobis: There are a lot of people trying to do this, but it won't work as a holistic approach to safety or efficacy. You are looking at just one target, and you can't assume that this will represent the whole organism. A liver cell will tell you something about liver toxicity, but it won't tell you anything about neurotoxicity or renal toxicity. It could be part of a tiered approach, but it is not the whole answer.

Azzi: How does your office work? We have found that a derivative of tocopherol, tocopherol phosphate, is present in cells. It is a natural compound that is more potent than the original tocopherol. How could I interact with your office in order to have support for further studies?

Coates: The ODS doesn't directly fund grants. We do co-fund grants in collaboration with the NIH Institutes and Centers. Grant applications come into the NIH, they are customarily reviewed centrally, and then assigned to an Institute or Center for funding. The Institute might bring an application to our attention: it might have a wonderful score and they would come to us and ask whether we are interested in co-funding it. One way that an investigator could enhance the chances of our co-funding would be to have a conversation with us in the beginning, letting us know that he/she is planning to submit the grant. We could monitor its progress and make the contact with the relevant Institute proactively.

FINAL DISCUSSION

Rayman: So far we haven't discussed ω-3 fatty acids (also known as *n*-3 fatty acids). These are important supplements in the UK, and represent a considerable proportion of money spent on supplements, especially among the elderly. This is an area that interests me, but it isn't my research field. In the UK, ω-3 fatty acids are mostly taken as cod liver oil, particularly by the over-60s. Do we need more of these? The ratio of fatty acids in our diet changed because of the evidence in the 1970s that linked coronary heart and cardiovascular disease to saturated fat intake. It also became known that polyunsaturated fatty acids appeared to reduce the risk of coronary heart diseases as well. We decreased our consumption of lard and stopped putting butter on our bread. Sunflower oil products began to take over. These are rich in ω-6 fatty acids. Another change has been that cows have shifted from being fed on grass to being fed on grain, so the ω-3 fatty acids in beef and milk have gone down while the ω-6 have increased. People are therefore consuming a lot more ω-6 fatty acids and fewer ω-3 fatty acids. In the USA, 13 g of linoleic acid (ω-6) is eaten per day but only 1.4 g of ω-3. In the UK our intake of linoleic acid has gone up from 10 g/day to 15 g/day, so our ratio is similar to that in the US. In countries where fish is eaten commonly, the ratio is better. People evolved to eat a ratio of about 2 or 3 to 1, so our current situation is very different. The most common ω-6 fatty acid in the diet is linoleic acid. This is converted through intermediates to arachidonic acid, which we also get directly from various foods. This conversion competes with that of the primary ω-3 fatty acid in the diet, α-linolenic acid, to eicosapentaenoic acid (EPA) and docosahexaenoic acid (DHA), which are the fatty acids associated with benefit. This is because the same enzymes are used. Thus if there is lots of linoleic acid in the diet, it will interfere with the generation of beneficial fatty acids. The incorporation of EPA into tissue phospholipids is therefore limited. EPA will give rise to prostaglandins and leukotrienes which are much less inflammatory than the series 2 prostaglandins and series 4 leukotrienes that come from arachidonic acid. It is said by people like Phil Calder that our diet now has a considerably greater inflammatory potential than was the case 30 years ago. Does the change in intake ratios matter? There is some evidence of benefit of ω-3 long chain polyunsaturated fatty acids (EPA and DHA) in inflammatory conditions such as rheumatoid arthritis (Calder & Zurier 2001). There are also effects on coronary heart disease (CHD) risk and ventricular arrhythmia (Brouwer et al 2006). There is also evidence for effects on depression and schizophrenia (Peet 2003). Work by Stuart Forsyth (Forsyth & Hornstra 2001) and others has shown the importance of ω-3 fatty acids on brain and visual development in children. Alex Richardson in Oxford has published on effects on attention deficit/hyperactivity disorder (ADHD) and dyspraxia (Richardson 2006). There is work on cod liver oil (Curtis et al 2002) showing a beneficial effect on cartilage integrity and there is some work on diabetic neuropathy (Horrobin 1997). So perhaps we do need to worry about this change in ratio and decreasing intake of ω-3 polyunsaturated fatty acids. However, in these different health conditions, sometimes the effects have been less strong than one might have expected from the biochemistry. This may lie in the fact that there are polymorphisms that affect the response to fish oil. This is particularly relevant to any condition where there is inflammation. We know tumour necrosis factor (TNF)-α production varies widely among individuals because of polymorphisms in the promoter regions of TNF-α and lymphotoxin-α. In response to an inflammatory stimulus, this influences the amount of TNF-α produced. The effects are complex, but for people who are in the highest tertile of TNF-α production, fish oil will inhibit its production. If you fall into the middle tertile, there is no effect, but if you are in the lowest tertile you don't want to be taking fish oil because this will stimulate TNF-α production. How do we redress the balance? We can decrease the amount of ω-6 fatty acids in our diet by changing our cooking oil to rapeseed or canola, or changing spreads to those which are rich in olive oil. We can increase the amounts of ω-3 by eating oily fish, but to achieve the anti-inflammatory effects you'd need six to seven servings a week of oily fish to get 3 g/day of EPA and DHA. This is the minimum level shown in trials to be beneficial in rheumatoid arthritis. So supplementation with fish body oil capsules is an option. This fish body oil has more EPA and DHA than cod liver oil. Liver oil has high levels of vitamin A and D, and you could overdose on this. Flaxseed oil/linseed oil have short-chain ω-3 fatty acids and there is no evidence that there is benefit from taking these. The conversion rates from the short chain to long chain forms are very low except in pregnant women. There are other potential benefits of taking supplements rather than eating oily fish: the supplements have lower levels of dioxins and mercury than fish. For me, the research question here is to what extent do the equivocal results in some research areas reflect the genetic homogeneity within studied populations?

Scott: I don't know whether the genotyping could be done in populations who are high fish oil consumers, but they presumably are polymorphic for these genotypes as well?

Rayman: I would guess so. The study I mentioned here was done by Bob Grimble in Southampton. This is the only study I know of that has looked at these polymorphisms.

Azzi: Is there a gradient, North–South, in relation to the diseases you mentioned?

Rayman: Not that I know of.

Azzi: The old study from Fred Gey showed that in the south of Europe there was very low cardiovascular disease and very high cardiovascular disease in the north (Gey et al 1991). Finland was an outlier. I think that in your case the geographic distribution goes in the opposite way.

Rayman: These are all multifactorial complex diseases. You can't expect simplistic answers.

Stocker: There was also the Lyon study from de Lorgeril (de Lorgeril et al 2005). This doesn't fit in with the results you mentioned about the ω-3 to ω-6 ratios.

Alexander: Has it been proven that it is the ratio that is important? When it comes to cancer there is some evidence that it is the absolute amount of ω-3 fatty acids that matters.

Rayman: Yes, you are right. But the ratio is an easy way to look at it, and also there is competition in the two metabolic pathways.

Azzi: As we come to the end of this meeting, it would be a good idea to try to summarize briefly what we have discussed, beginning with the important topic of biomarkers.

Aggett: I'd suggest that the key characteristics of biomarkers are pretty well accepted. There may be differences in terminology, and in some sectors people try to differentiate between biomarkers as factors if they can show that there is a causal relationship, and restrict their comments to calling them indicators if no more than an association has been demonstrated. There is a discipline emerging around biomarkers. Possibly the important message that we need to emphasize is that a lot of things that are called biomarkers do not meet the generally accepted criteria. This is a criticism that can be levelled at some areas of nutritional science. We need to get the culture of the discipline to appreciate this.

Boobis: I would add that we need to identify a mechanism to support the validation or demonstration of fitness for purpose of markers. This is a substantial undertaking. How do we identify a funding source? What tends to happen is that much of the work on biomarkers is being done by industry and they are using it to the extent that it is useful for their purposes. It may not be demonstrated to an external observer's satisfaction that this is a true biomarker.

Aggett: Often, things are being measured just because they can be measured. People aren't enquiring intelligently into the mechanism to come up with more credible markers.

Coates: It is nuts and bolts kind of stuff. There may or may not be a hypothesis in there: more often than not, it's just that we need these kinds of data. The NIH has joined up with the FDA and an industry consortium to form a biomarkers consortium. The idea is that there needs to be a concerted effort to get this information, for different reasons. It's in its early stages. This would get a lot of money from industry, put it into a pool, and administered by the NIH.

Scott: In an analytical laboratory being run as part of a clinical service there are good quality control systems. If you doing this in a research context there aren't any good standards. When CDC looked at some of the biomarkers in my own area, and they got the 20 supposedly best labs in the world to measure folate, the values were all over the place. It was scandalous. The same was found for B12. Many people doing these studies are not from a clinical pathology discipline and they don't understand the issues of quality control very well. This is a big issue.

Azzi: Is there any chance that the efficacy of compounds could be assessed by proteomics or metabolomics? Industry likes these kinds of systematic approaches. Is this a direction we should be taking?

Boobis: I think there is great potential. But as with other areas of the biomarker field, the early promise of a result is often acted on too quickly. No one demonstrates the robustness of the measurement. Nowhere is there more interlaboratory variability than in some of the platforms used to measure biomarkers. The interpretability of what makes a characteristic profile has yet to be determined. Almost no one has looked at the background variability. If you took 1000 people and did a metabonomic assessment of their urine over 24 h, how would it vary?

Przyrembel: I think it depends a lot on what you call efficacy. By one of these modern techniques, demonstrating systematic change doesn't mean efficacy; it may just mean a change.

Manach: There is great potential, but it is too early to say. One of the objectives is to find early biomarkers of the changes that will lead to the disease. We are a long way away from using these techniques.

Azzi: The next subject that requires some final comments is that of the risk assessment and claims.

Coates: We have learned that there are a lot of commonalities in the needs, whether the outcome is to inform risk assessment or inform labelling, or to inform the development of appropriate research agendas. There needs to be clear description of what the questions are at the beginning. These questions need to be refined as necessary along the way. This issue can complicate future policy implementation. My principal approach is to keep the analysis separate from the policy implications. The analysis is an important function of this, but every day I credit the people who have to do the tough job of implementing things. I don't think they make their job any easier if they drift back and forth between the analysis, policy and implementation.

Boobis: The idea of systematic evaluation of the data is important. There should be a clear statement of what we think we know and what we think we don't know.

Then the policy is based upon that conclusion. The risk assessor would conclude that they don't know about the safety of this material because they don't have the information. This is a data gap which may or may not be filled: it depends on whether the risk manager feels whether it is important or not. The same applies to efficacy: is there evidence for efficacy, and if there isn't it should be clearly stated that we just don't have the information.

Yetley: I have heard several times at this meeting that we need to differentiate between physiological effects and pharmacological effects. How can these concepts be operationalized?

Przyrembel: I agree. This is always used as a surrogate to differentiate between something that is not clearly a drug or a supplement. It doesn't help at all to call something pharmacological and something else physiological without a definition of the respective terms. Of course, 'pharmacological' in most cases is related to some pathophysiological process. But then you have to identify this process and be certain that it is beyond normal physiological processes, which also need to be characterised both with respect to mechanism and effective doses.

Azzi: AVED (ataxia with isolated vitamin E deficiency) is an example of a pathological process related to the lack of vitamin E retention. Supplementation with enough vitamin E to compensate for this would be a physiological intervention. But if vitamin E, at higher concentration, has a preventing effect on atherosclerosis, a higher physiological dose would be required to protect against this other disease. Thus the definition of 'physiological dose' depends on the chosen endpoint.

Przyrembel: It is relatively easy for the essential nutrients.

Taylor: At the FDA we have struggled with this idea of when a substance becomes pharmacological. The conclusion that dosing with 'more' of a substance makes it pharmacological rather than nutritional has not been supported by the experts we have consulted.

Boobis: My take is that it is to do with homeostasis. Physiological is within the homeostatic range of variation. Pharmacological goes beyond this. This is an area where genomics and proteomics will probably help us.

Azzi: In 1982 in this same place (it was then called the Ciba Foundation) a meeting on vitamin E took place. At that symposium, Dr Ingold stated that vitamin E is established as the best physiological antioxidant (Burton et al 1983). We heard at this meeting that Barry Halliwell no longer believes this, even though he used to. This is an example of how science progresses and how scientists have to respond to its changes.

Aggett: We have just discussed whether supplements are physiological or pharmacological. The important issue with selenium is that it is potentially correcting a deficiency. This adds a different prioritization. The data that Jan Alexander presented caused those of us involved with setting the WHO levels for selenium to consider the need to go back and look at these seriously.

Rayman: I would endorse this.

Azzi: Herbal medicine is a blooming topic, with all related problems.

Yong: Perhaps it is helpful to think about an analogy with wine, because getting a good quality herbal medicine is like how to get a good quality wine. First you must know the species and where it came from (for example, Château Latour), then it has to be stored correctly, and you have to know something about the chemistry, acidity, tannins, sugar, ethanol content, etc. But this is not sufficient: you need a biological response—you need a good nose to measure the final bouquet, the summated effect.

Azzi: Let me conclude with a little story told to me by Hugo Theorell, a Nobel Prize winner in Physiology or Medicine in 1955 'for his discoveries concerning the nature and mode of action of oxidation enzymes'. I was sitting near him at an official banquet in 1971 and to avoid an embarrassing silence I asked 'Professor Theorell, how did you manage to get a Nobel Prize?' He thought for a moment, and replied that there was a captain many years ago with his soldiers in the outskirts of Stockholm planning to do an exercise. But a lady came and told him that she had lost her precious ring. At this his point he decided that the exercise that day for his 100 soldiers was to search each of them a meter, for the length of the field. When the ring was found, the lady congratulated the soldier who had found it profusely. The soldier humbly argued that she should not thank him: without the other 99 soldiers the ring could not have been found. Each one of us is a soldier in the field of science: I hope that one of you is going to find the ring. But everyone's work is equally important to prepare the ground for the others.

References

Gey KF, Puska P, Jordan P, Moser UK 1991 Inverse correlation between plasma vitamin E and mortality from ischemic heart disease in cross-cultural epidemiology. Am J Clin Nutr 53:326S–334S

Brouwer IA, Geelen A, Katan MB 2006 n-3 fatty acids, cardiac arrhythmia and fatal coronary heart disease. Prog Lipid Res 45:357–367

Burton GW, Cheeseman KH, Doba T, Ingold KU, Slater TF 1983 Vitamin E as an antioxidant in vitro and in vivo. In: Biology of vitamin E. Elsevier/Excerpta Medica/North Holland, Amsterdam (Ciba Found Symp 101) p 4–18

Calder PC, Zurier RB 2001 Polyunsaturated fatty acids and rheumatoid arthritis. Curr Opin Clin Nutr Metab Care 4:115–121

Curtis CL, Rees SG, Little CB et al 2002 Pathological indicators of degradation and inflammation in human osteoarthritic cartilage are abrogated by exposure to n-3 fatty acids. Arthritis Rheum 46:1544–1553

de Lorgeril M, Salen P, Guiraud A, Zeghichi S, Boucher F, de Leiris J 2005 Lipid-lowering drugs and essential omega-6 and omega-3 fatty acids in patients with coronary heart disease. Nutr Metab Cardiovasc Dis 15:36–41

Forsyth S, Hornstra G 2001 Essential fatty acids. Maternal and infant nutrition. Pract Midwife 4:34–37

Horrobin DF 1997 Essential fatty acids in the management of impaired nerve function in diabetes. Diabetes 46:S90–S93

Peet M 2003 Eicosapentaenoic acid in the treatment of schizophrenia and depression: rationale and preliminary double-blind clinical trial results. Prostaglandins Leukot Essent Fatty Acids 69:477–485

Richardson A 2006 Omega-3 fatty acids in ADHD and related neurodevelopmental disorders. Int Rev Psychiatry 18:155–172

Index of Contributors

*New participating co-authors are indicated by asterisks. Entries in **bold** indicate papers; other entries refer to discussion contributions*

Subject Index